# Azure Networking Cookbook

## Second Edition

Practical recipes for secure network infrastructure, global application delivery, and accessible connectivity in Azure

Mustafa Toroman

# Azure Networking Cookbook, Second Edition

Author: Mustafa Toroman

Technical Reviewers: Kapil Bansal, Rithin Skaria

Managing Editors: Mamta Yadav, Siddhant Jain

Acquisitions Editors: Ben Renow-Clarke and Divya Mudaliar

Production Editor: Deepak Chavan

Editorial Board: Alex Patterson, Arijit Sarkar, Ben Renow-Clarke, Dominic Shakeshaft, Edward Doxey, Joanne Lovell, and Vishal Bodwani

First Published: March 2019

Second Published: October 2020

Production Reference: 1281020

ISBN: 978-1-80056-375-9

Published by Packt Publishing Ltd.

Livery Place, 35 Livery Street

Birmingham B3 2PB, UK

# Table of Contents

## Chapter 5: Local and virtual network gateways     75

## Chapter 6: DNS and routing         85

## Chapter 9: Connecting to resources securely     149

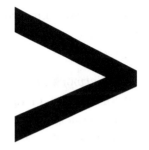

# Preface

**About**

This section briefly introduces the author and technical reviewers, the coverage of this cookbook, the technical skills you'll need to get started, and the hardware and software required to complete all of the included recipes.

## About Azure Networking Cookbook, Second Edition

Azure's networking services enable organizations to manage their networks effectively. Azure paves the way for an enterprise to achieve reliable performance and secure connectivity.

*Azure Networking Cookbook, Second Edition* starts with an introduction to Azure networking, covering basic steps such as creating Azure virtual networks, designing address spaces, and creating subnets. You will go on to learn how to create and manage network security groups, application security groups, and IP addresses in Azure.

As you progress, you will explore various aspects such as Site-to-Site, Point-to-Site, and virtual network–to–virtual network connections, DNS and routing, load balancers, and Traffic Manager. This cookbook covers every aspect and function you need to be aware of, providing practical recipes to help you go from having a basic understanding of cloud networking practices to being able to plan, implement, and secure your network infrastructure with Azure.

This cookbook will not only help you upscale your current environment but also instruct you on how to monitor, diagnose, and ensure secure connectivity. After learning how to create a robust environment, you will gain meaningful insights from recipes on best practices.

By the end of this cookbook, you will possess sufficient practical experience in providing cost-effective solutions to facilitate efficient connectivity in your organization.

## About the author

**Mustafa Toroman** is a solution architect with Authority Partners. With years of experience in designing and monitoring infrastructure solutions, lately, he has been focusing on designing new solutions in the cloud and migrating existing solutions to the cloud. He is very interested in DevOps processes, and he's also an Infrastructure as Code enthusiast. Mustafa has over 50 Microsoft certifications and has been a Microsoft Certified Trainer since 2012. He often speaks at international conferences about cloud technologies, and he has been awarded the MVP award for Azure for the last five years in a row.

Mustafa also authored *Hands-On Cloud Administration in Azure* and co-authored *Learn Node.js with Azure* and *Mastering Azure Security*, all published by Packt.

# About the reviewers

**Kapil Bansal** is a lead DevOps engineer at S&P Global Market Intelligence, India. He has more than 12 years of experience in the IT industry, having worked on Azure cloud computing (PaaS, IaaS, and SaaS), Azure Stack, DevSecOps, Kubernetes, Terraform, Office 365, SharePoint, release management, application lifecycle management (ALM), Information Technology Infrastructure Library (ITIL), and Six Sigma. He has worked with companies such as IBM India Pvt Ltd, HCL Technologies, NIIT Technologies, Encore Capital Group, and Xavient Software Solutions, Noida, and has served multiple clients based in the United States, the UK, and Africa, such as T-Mobile, World Bank Group, H&M, WBMI, Encore Capital, and Bharti Airtel (India and Africa). Kapil has also reviewed *Hands on Kubernetes on Azure and Azure Networking Cookbook* published by Packt. Additionally, he has contributed in *Practical Microsoft Azure IaaS* and *Beginning SharePoint Communication Sites* published by Apress.

**Rithin Skaria** is an open-source evangelist with over 7 years of experience managing open-source workloads in Azure, AWS, and OpenStack. He is currently working for Microsoft and is a part of several open-source community activities being conducted within Microsoft. He is a Microsoft Certified Trainer, Linux Foundation Certified Engineer and Administrator, Kubernetes Application Developer and Administrator, and also a Certified OpenStack Administrator. When it comes to Azure, he has four certifications, including for solution architecture, Azure administration, DevOps, and security, and he is also certified in Microsoft 365 Administration. He has played a vital role in several open-source deployments and the administration and migration of these workloads to cloud. He co-authored *Linux Administration on Azure* and *Azure for Architects - Third Edition* published by Packt.

## Learning objectives

By the end of this cookbook, you will be able to:

- Create Azure networking services.
- Create and work on hybrid connections.
- Configure and manage Azure networking services.
- Design high-availability network solutions in Azure.
- Monitor and troubleshoot Azure networking resources.
- Use different methods of connecting local networks to Azure virtual networks.
- Use different methods to secure networks.

## Audience

This cookbook is targeted at cloud architects, cloud solution providers, or any stakeholders dealing with networking on Azure. Basic familiarity with Azure would be a plus.

## Approach

*Azure Networking Cookbook, Second Edition* achieves an ideal blend of theory and hands-on training to help you prepare for the real-world connectivity challenges faced by enterprises.

## To get the most out of this book

This book assumes a basic level of knowledge of cloud computing and Azure. To use this book, all you need is a valid Azure subscription and internet connectivity. A Windows 10 machine with 4 GB of RAM is sufficient for using PowerShell.

## Hardware requirements

The Azure portal is a web-based console that runs on all modern browsers for desktops, tablets, and mobile devices. To use the Azure portal, you must have JavaScript enabled on your browser.

## Software requirements

We recommend that you use the most up-to-date browser that's compatible with your operating system. The following browsers are supported:

- Microsoft Edge (latest version)
- Internet Explorer 11
- Safari (latest version, Mac only)
- Chrome (latest version)
- Firefox (latest version)

# Conventions

Code words in the text, folder names, filenames, file extensions, pathnames, dummy URLs, and user input, are shown as follows:

"Furthermore, we can use additional switches, such as **-SKU** for selecting **Basic** or **Standard**, **-IPAddressVersion** for choosing between IPv4 and IPv6, and **-DomainNamelabel** to specify the DNS label."

A block of code is set as follows:

```
$VirtualNetwork = Get-AzVirtualNetwork -Name 'Packt-Script' '
-ResourceGroupName 'Packt-Networking-Script'
Add-AzVirtualNetworkSubnetConfig -Name BackEnd '
-AddressPrefix 10.11.1.0/24 '
-VirtualNetwork $VirtualNetwork
$VirtualNetwork | Set-AzVirtualNetwork
```

# Download Resources

The code bundle for this book is hosted on GitHub at https://github.com/PacktPublishing/Azure-Networking-Cookbook-Second-Edition. You can find the files used in this book, which are referred to at relevant instances. We also have other code bundles from our rich catalog of books and videos available at https://github.com/PacktPublishing/. Check them out!

# 1

# Azure Virtual Network

In this very first chapter, we will learn about the basics of Azure networking, including creating Azure virtual networks and designing address spaces and subnets. This will lay the foundation for all future recipes that will be covered in this chapter.

We will cover the following recipes in this chapter:

- Creating a virtual network in the Azure portal
- Creating a virtual network with PowerShell
- Adding a subnet in the Azure portal
- Adding a subnet with PowerShell
- Changing the address space size
- Changing the subnet size

## Technical requirements

For this chapter, the following is required:

- An Azure subscription
- Azure PowerShell

The code samples can be found at https://github.com/PacktPublishing/Azure-Networking-Cookbook-Second-Edition/tree/master/Chapter01.

# Creating a virtual network in the Azure portal

Azure Virtual Network represents your local network in the cloud. It enables other Azure resources to communicate over a secure private network without exposing endpoints over the internet.

## Getting ready

Before you start, open a web browser and go to the Azure portal at https://portal.azure.com.

## How to do it...

In order to create a new virtual network using the Azure portal, take the following steps:

1. In the Azure portal, select **Create a resource** and choose **Virtual network** under **Networking** (or search for `virtual network` in the search bar). A new pane will open, where we need to provide information for the virtual network. First, select the **Subscription** option we want to use and the **Resource group** option for where the virtual network will be deployed. Then, include a name and select a region (of the Azure datacenter) for where the virtual network will be deployed. An example is shown in *Figure 1.1*:

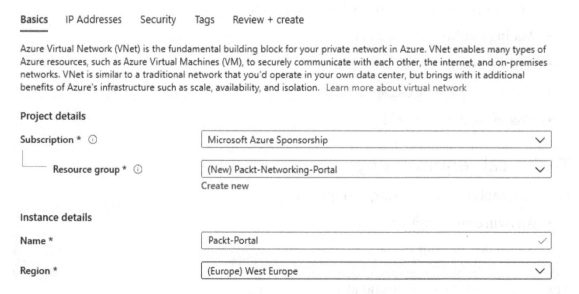

Figure 1.1: Creating an Azure virtual network

2. In the next pane, we first need to define the address space and define the **Subnet name** and **Subnet address range** values for the first subnet. After the address space is defined, as shown in *Figure 1.2*, we will receive a message stating that **This virtual network doesn't have any subnets**. Therefore, we need to select the **Add subnet** option:

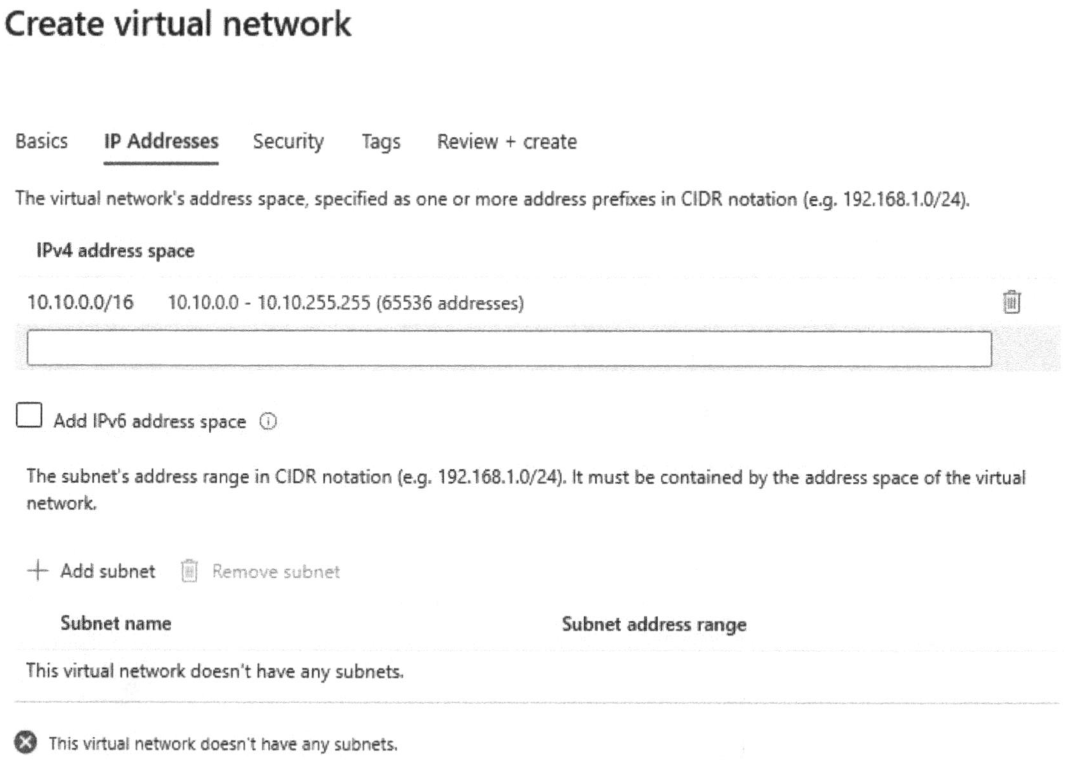

## Create virtual network

Basics   **IP Addresses**   Security   Tags   Review + create

The virtual network's address space, specified as one or more address prefixes in CIDR notation (e.g. 192.168.1.0/24).

**IPv4 address space**

10.10.0.0/16   10.10.0.0 - 10.10.255.255 (65536 addresses)     🗑

☐ Add IPv6 address space ⓘ

The subnet's address range in CIDR notation (e.g. 192.168.1.0/24). It must be contained by the address space of the virtual network.

+ Add subnet   🗑 Remove subnet

| Subnet name | Subnet address range |
| --- | --- |
| This virtual network doesn't have any subnets. | |

❌ This virtual network doesn't have any subnets.

**Figure 1.2: Configuring a virtual network address space and subnet**

3. In the **Add subnet** pane, we need to define **Subnet name** and **Subnet address range**. Optionally, we can add service endpoints we want to connect to the virtual network. Service endpoints allow us to connect to Azure services in a secure way, over Azure backbone infrastructure, without needing a public IP address. An example is shown in *Figure 1.3*:

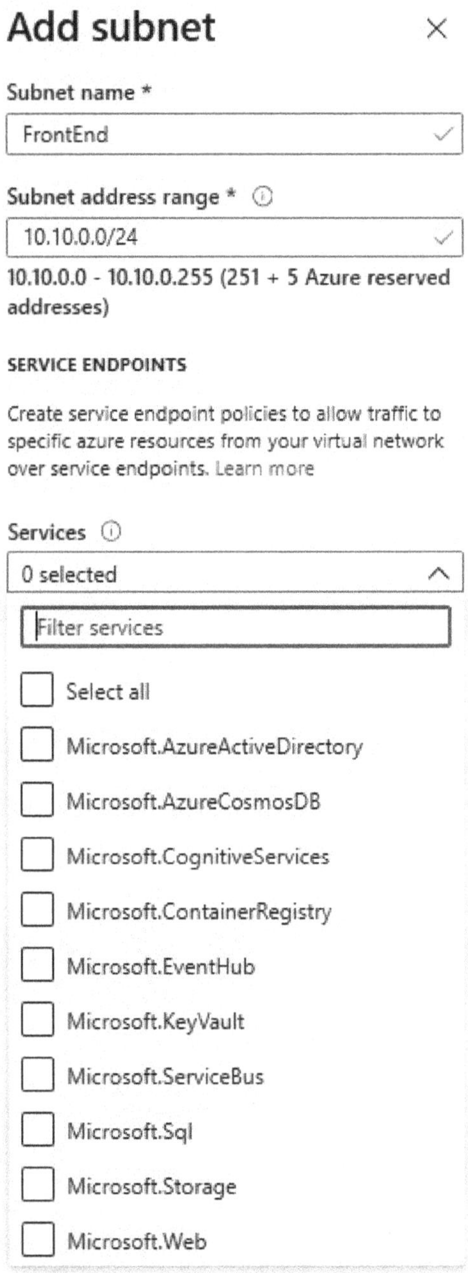

## Add subnet                                                     ✕

Subnet name *

| FrontEnd                                                   ✓ |

Subnet address range *   ⓘ

| 10.10.0.0/24                                               ✓ |

10.10.0.0 - 10.10.0.255 (251 + 5 Azure reserved addresses)

**SERVICE ENDPOINTS**

Create service endpoint policies to allow traffic to specific azure resources from your virtual network over service endpoints. Learn more

Services ⓘ

| 0 selected                                                 ⌃ |

| Filter services |

☐ Select all

☐ Microsoft.AzureActiveDirectory

☐ Microsoft.AzureCosmosDB

☐ Microsoft.CognitiveServices

☐ Microsoft.ContainerRegistry

☐ Microsoft.EventHub

☐ Microsoft.KeyVault

☐ Microsoft.ServiceBus

☐ Microsoft.Sql

☐ Microsoft.Storage

☐ Microsoft.Web

Figure 1.3: Adding a subnet

4. After we have added the first subnet, in our case, **FrontEnd**, we can add more subnets to the virtual network or proceed to the **Security** section, as shown in *Figure 1.4*:

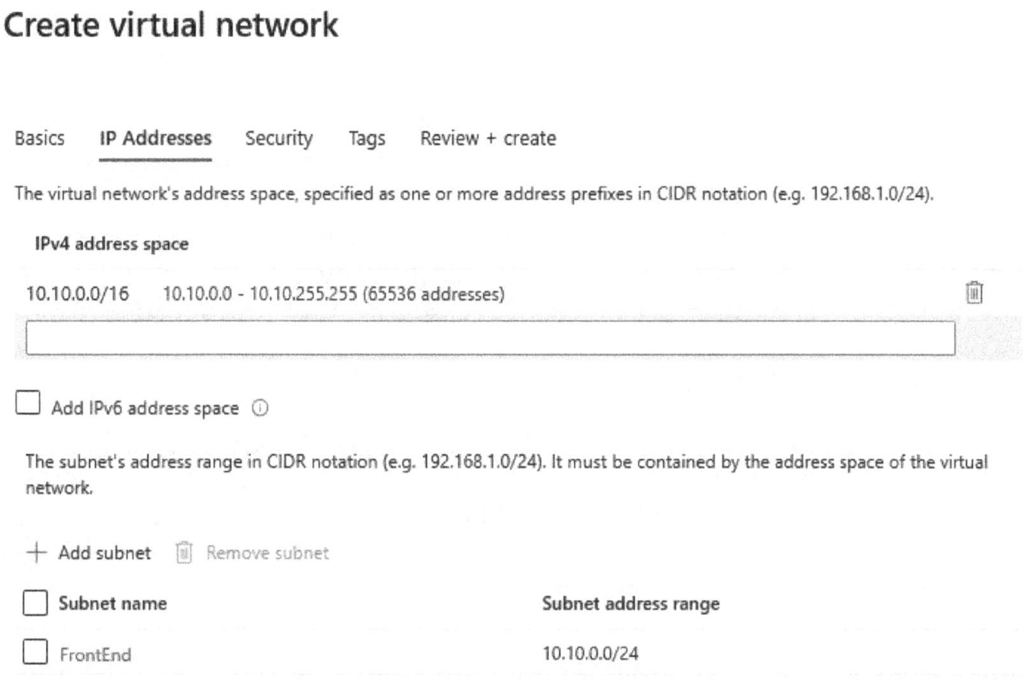

## Create virtual network

Basics    **IP Addresses**    Security    Tags    Review + create

The virtual network's address space, specified as one or more address prefixes in CIDR notation (e.g. 192.168.1.0/24).

**IPv4 address space**

10.10.0.0/16    10.10.0.0 - 10.10.255.255 (65536 addresses)

☐ Add IPv6 address space ⓘ

The subnet's address range in CIDR notation (e.g. 192.168.1.0/24). It must be contained by the address space of the virtual network.

＋ **Add subnet**    🗑 Remove subnet

| ☐ **Subnet name** | **Subnet address range** |
|---|---|
| ☐ FrontEnd | 10.10.0.0/24 |

Figure 1.4: Adding the FrontEnd subnet

5. In the **Security** section, we can choose whether we want to enable **Bastion Host**, **DDoS protection**, and **Firewall**. If any of these options are enabled, we need to provide additional information for that service. Afterward, we can optionally add tags, or skip that and create the service. An example is shown in *Figure 1.5*:

## Create virtual network

Basics    IP Addresses    Security    Tags    Review + create

BastionHost ⓘ    ( Disabled ) Enabled

DDoS protection ⓘ    ( Basic ) Standard

Firewall ⓘ    ( Disabled ) Enabled

Figure 1.5: Toggling security options

6. Creating a virtual network usually does not take much time and should be completed in under two minutes. Once the deployment is finished, we can start using the virtual network.

## How it works...

We deploy virtual networks to **Resource group** under **Subscription** in the Azure datacenter that we choose. **Region** and **Subscription** are important parameters; we will only be able to attach Azure resources to this virtual network if they are in the same subscription and region as the Azure datacenter. The address space option defines the number of IP addresses that will be available for our network. It uses the **Classless Inter-Domain Routing (CIDR)** format and the largest range we can choose is **/8**. In the portal, we need to create an initial subnet and define the subnet address range. The smallest subnet allowed is **/29** and the largest is **/8** (however, this cannot be larger than the virtual network range). For reference, the range **10.0.0.0/8** (in CIDR format) will create an address range of **167772115** IP addresses (from **10.0.0.0** to **10.255.255.255**) and **10.0.0.0/29** will create a range of **8** IP addresses (from **10.0.0.0** to **10.0.0.7**).

# Creating a virtual network with PowerShell

PowerShell is a command-line shell and scripting language based on .NET Framework. It's often used by system administrators to automate tasks and manage operating systems. Azure PowerShell **Az** is a PowerShell module that allows us to automate and manage Azure resources. **Az** is also very often used to automate deployment tasks and can also be used to deploy a new Azure virtual network.

## Getting ready

Before we start, we need to make sure that we have the latest **Az** modules installed. To install **Az** modules, we need to run this command in the PowerShell console:

```
Install-Module -Name Az -AllowClobber -Scope CurrentUser
```

For more information, you can visit https://docs.microsoft.com/powershell/azure/install-az-ps?view=azps-4.5.0.

Before we start, we need to connect to the Azure subscription from a PowerShell console. Here's the command to do this:

```
Connect-AzAccountAzAccount
```

This will open a pop-up window where we need to input the credentials for the Azure subscription.

Afterward, we need to create a resource group where our virtual network will be deployed:

```
New-AzResourceGroup -name 'Packt-Networking-Script' -Location 'westeurope'
```

The output should be similar to that shown in *Figure 1.6*:

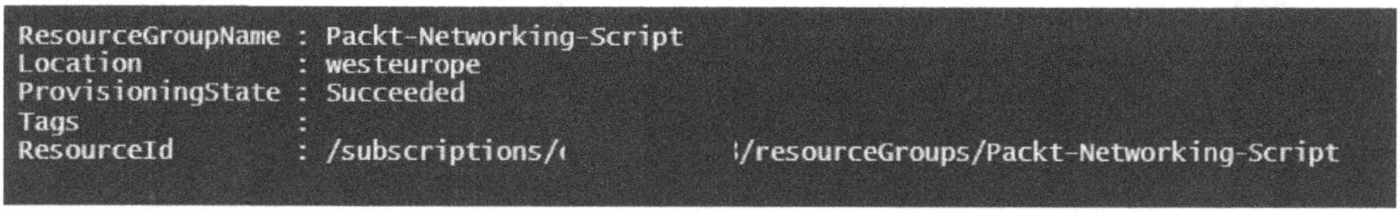

```
ResourceGroupName : Packt-Networking-Script
Location          : westeurope
ProvisioningState : Succeeded
Tags              :
ResourceId        : /subscriptions/(              :/resourceGroups/Packt-Networking-Script
```

Figure 1.6: Connecting to an Azure subscription from PowerShell

## How to do it...

Deploying an Azure virtual network is done in a single script. We need to define the parameters for the name, resource group, location, and address range. Here is an example script:

```
New-AzVirtualNetwork -ResourceGroupName 'Packt-Networking-Script' -Location
'westeurope' -Name 'Packt-Script' -AddressPrefix 10.11.0.0/16
```

You should receive the following output:

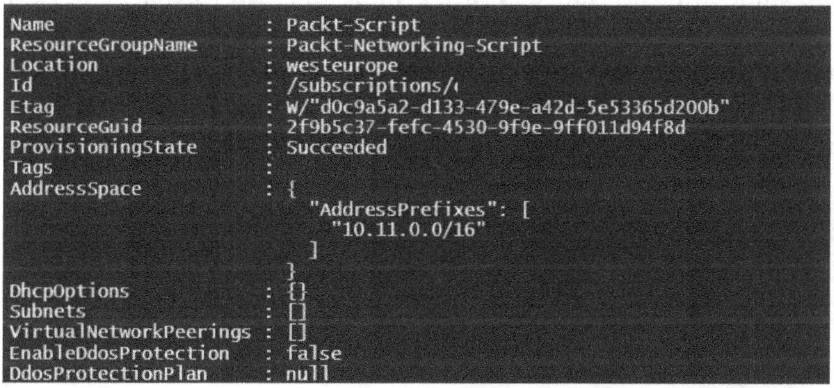

```
Name                   : Packt-Script
ResourceGroupName      : Packt-Networking-Script
Location               : westeurope
Id                     : /subscriptions/(
Etag                   : W/"d0c9a5a2-d133-479e-a42d-5e53365d200b"
ResourceGuid           : 2f9b5c37-fefc-4530-9f9e-9ff011d94f8d
ProvisioningState      : Succeeded
Tags                   :
AddressSpace           : {
                             "AddressPrefixes": [
                               "10.11.0.0/16"
                             ]
                         }
DhcpOptions            : {}
Subnets                : []
VirtualNetworkPeerings : []
EnableDdosProtection   : false
DdosProtectionPlan     : null
```

Figure 1.7: Deploying an Azure virtual network using a script

## How it works...

The difference between deploying a virtual network from the portal and using PowerShell is that no subnet needs to be defined in PowerShell. The subnet is deployed in a separate command that can be executed either when you are deploying a virtual network, or later on. We are going to see this command in the *Adding a subnet with PowerShell* recipe later in this chapter.

## Adding a subnet in the Azure portal

In addition to adding subnets while creating a virtual network, we can add additional subnets to our network at any time.

### Getting ready

Before you start, open a web browser and go to the Azure portal at https://portal.azure. com. Here, locate the previously created virtual network.

### How to do it...

In order to add a subnet to a virtual network using the Azure portal, we must take the following steps:

1.  In the **Virtual network** pane, go to the **Subnets** section.

2.  Select the **Add subnet** option.

3.  A new pane will open. We need to provide information for the subnet, including the **Name** value and the **Address range** value in CIDR format. The **Address range** value must be in the range limit of the virtual network address range and cannot overlap with the address range of other subnets in the virtual network. Optionally, we can add information for **Network security group**, **Route table**, **Service endpoints**, and **Subnet delegation**. These options will be covered in later recipes:

# Add subnet

Packt-Portal

Name *

BackEnd

Address range (CIDR block) *  ⓘ

10.10.1.0/24

10.10.1.0 - 10.10.1.255 (251 + 5 Azure reserved addresses)

NAT gateway ⓘ

None

☐ Add IPv6 address space

Network security group

None

Route table

None

Service endpoints

Services ⓘ

0 selected

Subnet delegation

Delegate subnet to a service ⓘ

None

**Figure 1.8: Adding the address range**

4. We can also add a gateway subnet in the same pane. To add a gateway subnet, select the **Gateway subnet** option.

For a gateway subnet, the only parameter we need to define is **Address range**. The same rules apply as for adding a regular subnet. This time, we don't have to provide a name, as it's already defined. You can add only one gateway subnet per virtual network. Service endpoints are not allowed in the gateway subnet:

Figure 1.9: Adding a gateway subnet for a virtual network

5. After the subnets are added, we can see the newly created subnets in the **Subnets** pane under the virtual network:

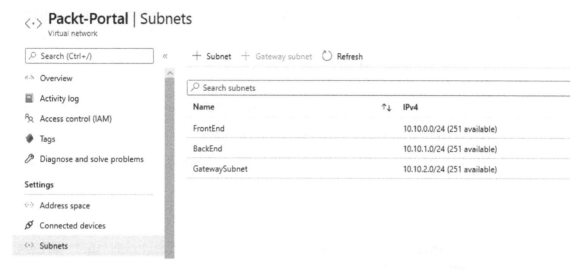

Figure 1.10: Viewing newly created subnets in the Subnets pane

## How it works...

A single virtual network can have multiple subnets defined. Subnets cannot overlap and must be in the range of the virtual network address range. For each subnet, four IP addresses are saved for Azure management and cannot be used. Depending on the network settings, we can define the communication rules between subnets in the virtual network. A gateway subnet is used for **Virtual Private Network (VPN)** connections, and this will be covered later in the cookbook.

Now, let's learn how to add a subnet using PowerShell.

# Adding a subnet with PowerShell

When creating an Azure virtual network with PowerShell, a subnet is not created in the same step and requires an additional command to be executed separately.

## Getting ready

Before creating a subnet, we need to collect information about the virtual network that the new subnet will be associated with. The parameters that need to be provided are the name of the virtual network and the resource group that the virtual network is located in:

```
$VirtualNetwork = Get-AzVirtualNetwork -Name 'Packt-Script'
-ResourceGroupName 'Packt-Networking-Script'
```

## How to do it...

1.  To add a subnet to the virtual network using PowerShell, we need to execute a command and provide the name and address prefix. The address prefix is again in CIDR format:

    ```
    Add-AzVirtualNetworkSubnetConfig -Name FrontEnd -AddressPrefix 10.11.0.0/24
    -VirtualNetwork $VirtualNetwork
    ```

2.  We need to confirm these changes by executing the following command:

    ```
    $VirtualNetwork | Set-AzVirtualNetwork
    ```

3.  We can add an additional subnet by running all commands in a single step, as follows:

    ```
    $VirtualNetwork = Get-AzVirtualNetwork -Name 'Packt-Script'
    -ResourceGroupName 'Packt-Networking-Script'
    Add-AzVirtualNetworkSubnetConfig -Name BackEnd -AddressPrefix 10.11.1.0/24
    -VirtualNetwork $VirtualNetwork
    $VirtualNetwork | Set-AzVirtualNetwork
    ```

## How it works...

The subnet is created and added to the virtual network, but we need to confirm the changes before they can become effective. When it comes to size, all the rules for creating or adding a subnet using the Azure portal, apply here as well; the subnet must be within the virtual network's address space and cannot overlap with other subnets in the virtual network. The smallest subnet allowed is **/29**, and the largest is **/8**, provided the value is within the virtual network's address space. For example, if you are creating a **/16** network, the largest value for the subnet will be **/16** only, as we cannot include a **/8** subnet in a **/16** address space.

## There's more...

We can create and add multiple subnets with a single script, as follows:

```
$VirtualNetwork = Get-AzVirtualNetwork -Name 'Packt-Script'
-ResourceGroupName 'Packt-Networking-Script'

$FrontEnd = Add-AzVirtualNetworkSubnetConfig -Name FrontEnd -AddressPrefix
10.11.0.0/24 -VirtualNetwork $VirtualNetwork

$BackEnd = Add-AzVirtualNetworkSubnetConfig -Name BackEnd -AddressPrefix
10.11.1.0/24 -VirtualNetwork $VirtualNetwork

$VirtualNetwork | Set-AzVirtualNetwork
```

# Changing the address space size

After the initial address space is defined during the creation of a virtual network, we can still change the address space size as needed. We can either increase or decrease the size of the address space or change the address space completely by using a new address range.

## Getting ready

Before you start, open a web browser and go to the Azure portal at https://portal.azure. com.

## How to do it...

In order to change the address space size for a virtual network using the Azure portal, we must observe the following steps:

1.  In the **Virtual network** pane, locate **Address space** under **Settings**.

2.  Next, click on **Address space** and change the value to the desired range. An example is shown in *Figure 1.11*:

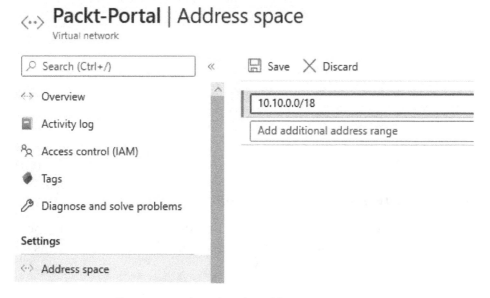

Figure 1.11: Changing the address space range

3.  After you have entered a new value for **Address space**, click **Save** to apply the changes.

## How it works...

Although you can change the address space at any time, there are some rules that determine what you can and cannot do. The address space cannot be decreased if you have subnets defined in the address space that would not be covered by the new address space. For example, if the address space were in the range of **10.0.0.0/16**, it would cover addresses from **10.0.0.1** to **10.0.255.254**. If one of the subnets was defined as **10.0.255.0/24**, we wouldn't be able to change the virtual network to **10.0.0.0/17**, as this would leave the subnet outside the new space.

The address space can't be changed to a new address space if you have subnets defined. In order to completely change the address space, you need to remove all subnets first. For example, if we had the address space defined as **10.0.0.0/16**, we would not be able to change it to **10.1.0.0/16**, since having any subnets in the old space would leave them in an undefined address range.

Let's see how to change the size of the newly created subnets.

# Changing the subnet size

Similar to the virtual network address space, we can change the size of a subnet at any time.

## Getting ready

Before you start, open a web browser and go to the Azure portal at https://portal.azure.com.

## How to do it...

In order to change the subnet size using the Azure portal, we must take the following steps:

1. In the **Virtual network** pane, select the **Subnets** option.

2. Select the subnet you want to change. In the **Subnets** option, enter a new value for the subnet size under **Address range**. An example of how to do this is shown in *Figure 1.12*:

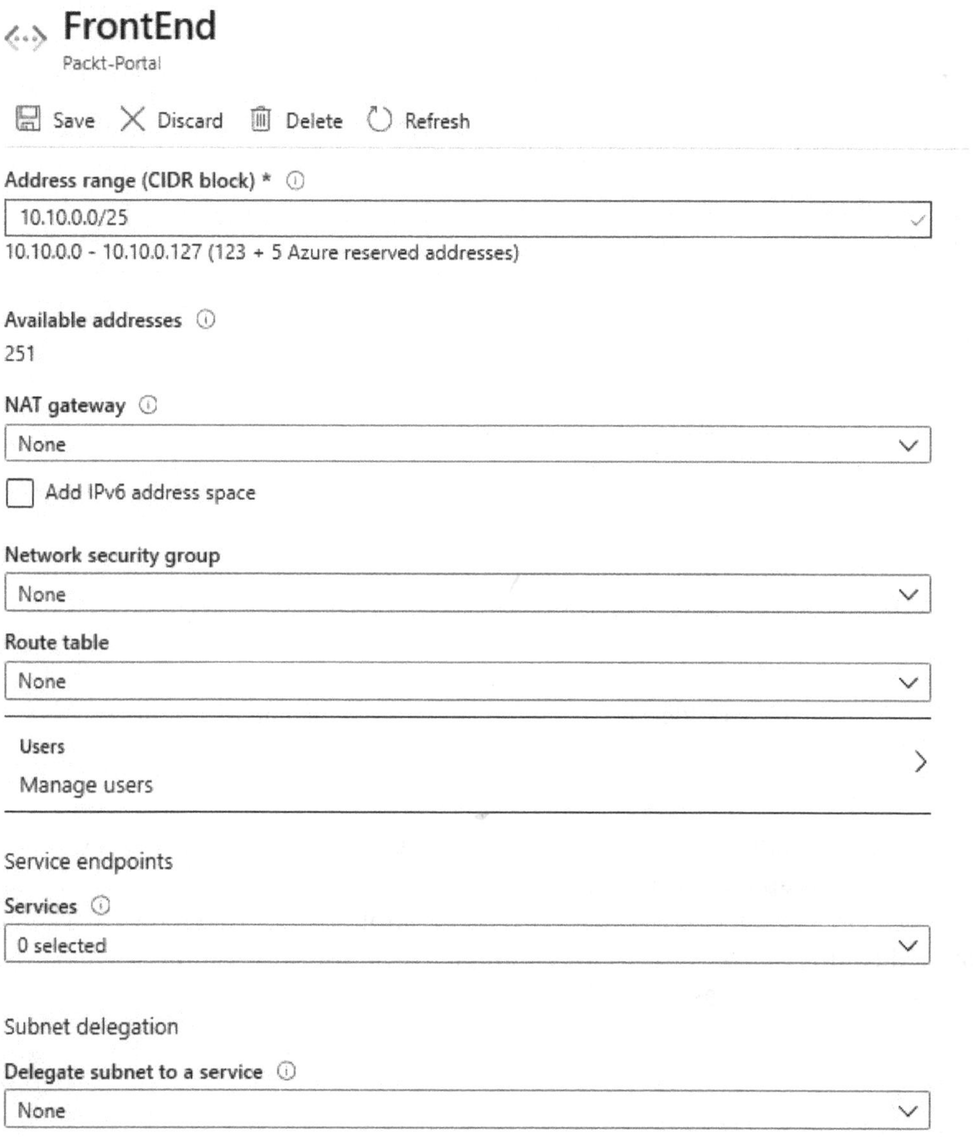

‹··› **FrontEnd**
Packt-Portal

💾 Save ✕ Discard 🗑 Delete ↻ Refresh

**Address range (CIDR block) * ⓘ**

| 10.10.0.0/25 | ✓ |

10.10.0.0 - 10.10.0.127 (123 + 5 Azure reserved addresses)

**Available addresses** ⓘ

251

**NAT gateway** ⓘ

| None | ⌄ |

☐ Add IPv6 address space

**Network security group**

| None | ⌄ |

**Route table**

| None | ⌄ |

Users
Manage users                                                    ›

Service endpoints

**Services** ⓘ

| 0 selected | ⌄ |

Subnet delegation

**Delegate subnet to a service** ⓘ

| None | ⌄ |

Figure 1.12: Changing the subnet size using the Azure portal

3. After entering a new address range value, click **Save**.

4. In the **Subnets** list, you can see that the changes have been applied and the address space has changed, as shown in *Figure 1.13*:

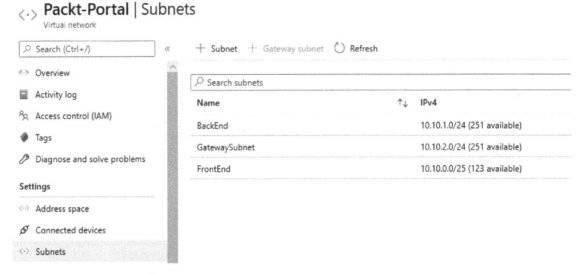

**Figure 1.13: Viewing changes made in the subnet address range**

## How it works...

When changing the subnet size, there are some rules that must be followed. We cannot change the address space if it is not within the virtual network address space range, and the subnet range cannot overlap with other subnets in a virtual network. If devices are assigned to this subnet, we cannot change the subnet to exclude the addresses that these devices are already assigned to.

# 2

# Virtual machine networking

In this chapter, we'll cover Azure **Virtual Machines (VMs)** and the **network interface (NIC)** that is used as an interconnection between Azure VMs and Azure Virtual Network.

We will cover the following recipes in this chapter:

- Creating Azure VMs
- Viewing VM network settings
- Creating a new NIC
- Attaching an NIC to a VM
- Detaching an NIC from a VM

## Technical requirements

For this chapter, the following is required:

- An Azure subscription

# Creating Azure VMs

Azure VMs depend on virtual networking, and during the creation process, we need to define the network settings.

## Getting ready

Before we start, open a web browser and go to the Azure portal at https://portal.azure.com.

## How to do it...

In order to create a new VM using the Azure portal, we must use the following steps:

1. In the Azure portal, select **Create a resource** and choose the **Windows Server 2016 Datacenter** VM (or search for any VM image by searching for **image** in the **Search the Marketplace** search bar).

2. In the **Create a virtual machine** pane, we need to provide information for various options; not all of these are related to networking. First, we need to provide information on our Azure **Subscription** and **Resource group** (create a new resource group or provide an existing one).

3. In **Instance details**, we need to provide information for the **Virtual machine name**, **Region**, **Availability options**, and **Image** fields (for the **Image** field, leave the default or change to a different image from the drop-down menu). Some example settings are shown in *Figure* 2.1:

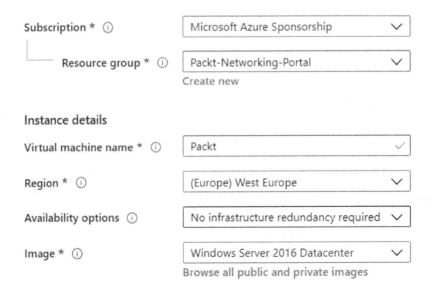

**Figure 2.1: Providing information for Instance details**

4.  Next, we need to select whether we want to use **Azure Spot instance** (where the VM runs on unused datacenter capacity at a lower price but can be turned off if resources are needed elsewhere) and provide information on our VM's **Size**, **Username**, and **Password**. Note that for **Username**, you can't use names such as admin, administrator, sysadmin, or root. The password must be at least 12 characters long and satisfy three of the four common rules (that is, having uppercase letters, lowercase letters, special characters, and numbers). An example of the completed screen is shown in *Figure 2.2*:

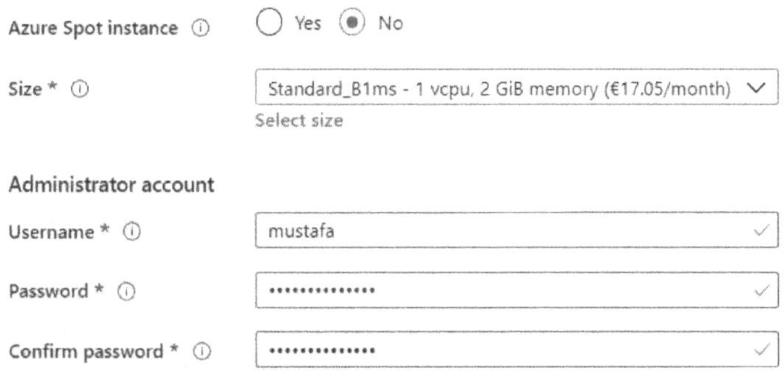

Azure Spot instance ⓘ   ◯ Yes  ⦿ No

Size * ⓘ   Standard_B1ms - 1 vcpu, 2 GiB memory (€17.05/month) ∨
Select size

Administrator account

Username * ⓘ   mustafa  ✓

Password * ⓘ   ••••••••••••••  ✓

Confirm password * ⓘ   ••••••••••••••  ✓

Figure 2.2: Configuring Azure Spot instance

5.  Next, we arrive at an option that concerns networking. We need to define whether we are going to allow any type of connection over a public IP address. We can select whether we want to deny all access or allow a specific port. Optionally, we can use **Hybrid Benefit** to use an existing license to save on costs. In the following example, I'm choosing **RDP (3389)**, but the dropdown also offers options for **SSH (22)**, **HTTP (80)**, and **HTTPS (443)**:

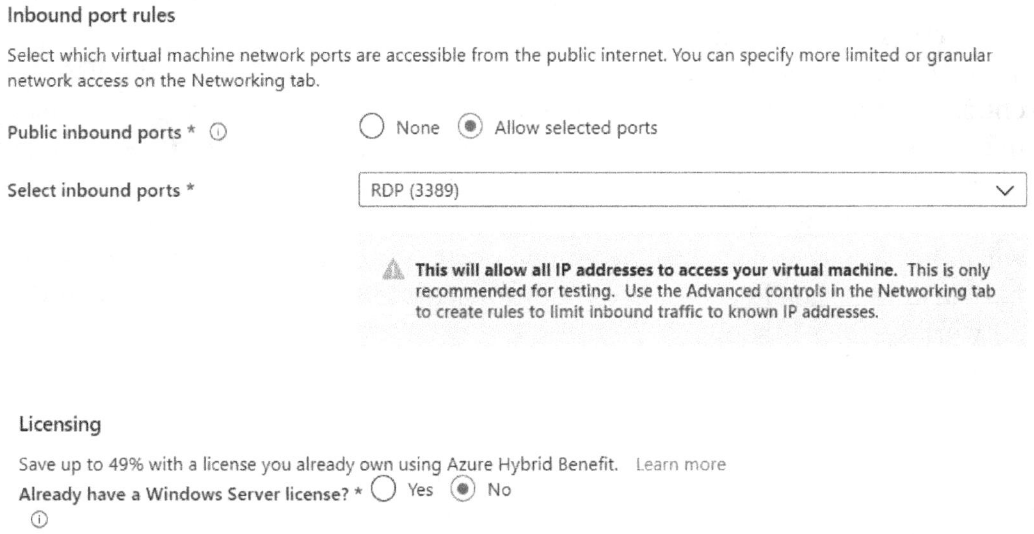

Inbound port rules

Select which virtual machine network ports are accessible from the public internet. You can specify more limited or granular network access on the Networking tab.

Public inbound ports * ⓘ   ◯ None  ⦿ Allow selected ports

Select inbound ports *   RDP (3389)  ∨

⚠ **This will allow all IP addresses to access your virtual machine.** This is only recommended for testing. Use the Advanced controls in the Networking tab to create rules to limit inbound traffic to known IP addresses.

Licensing

Save up to 49% with a license you already own using Azure Hybrid Benefit.   Learn more

**Already have a Windows Server license? *** ◯ Yes  ⦿ No
ⓘ

Figure 2.3: Defining inbound port rules

6. In the next section, we need to define disks. We can choose between **Premium SSD**, **Standard SSD**, and **Standard HDD**. An OS disk is required and must be defined. We can attach additional data disks as needed. Disks can be added at a later time, as well. The default encryption option is to use platform-managed keys, but we can select customer-managed keys if needed. An example of disk settings with only the OS disk is shown in *Figure 2.4*:

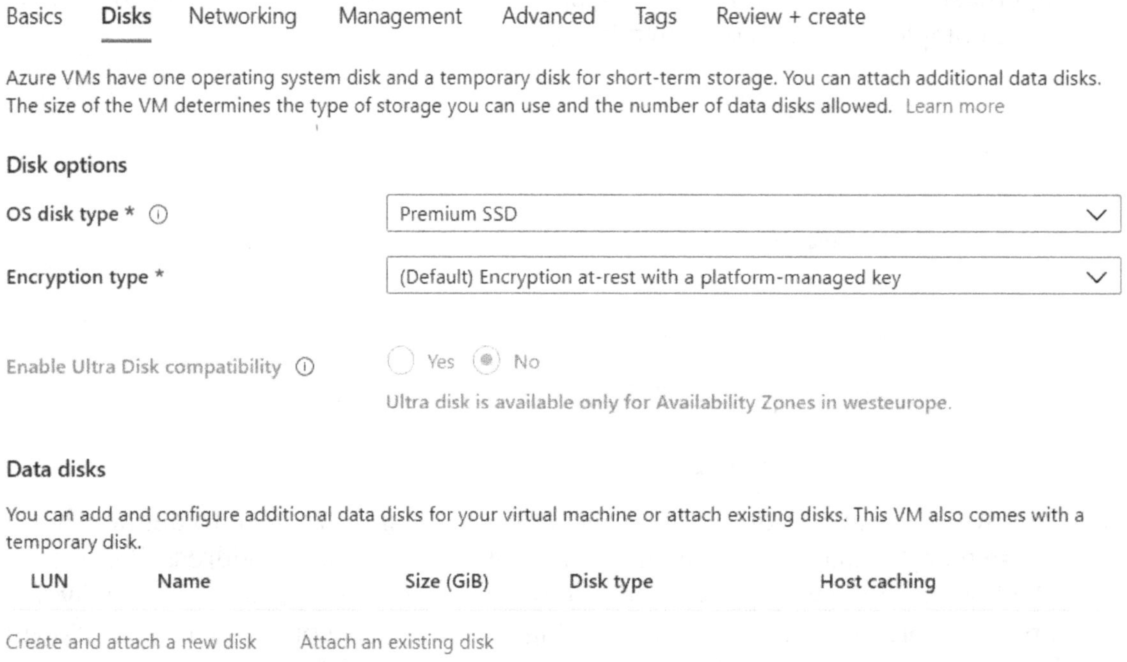

Basics    **Disks**    Networking    Management    Advanced    Tags    Review + create

Azure VMs have one operating system disk and a temporary disk for short-term storage. You can attach additional data disks. The size of the VM determines the type of storage you can use and the number of data disks allowed.  Learn more

**Disk options**

| OS disk type * ⓘ | Premium SSD | ⌄ |
|---|---|---|

| Encryption type * | (Default) Encryption at-rest with a platform-managed key | ⌄ |
|---|---|---|

Enable Ultra Disk compatibility ⓘ    ◯ Yes  ⦿ No

Ultra disk is available only for Availability Zones in westeurope.

**Data disks**

You can add and configure additional data disks for your virtual machine or attach existing disks. This VM also comes with a temporary disk.

| LUN | Name | Size (GiB) | Disk type | Host caching |
|---|---|---|---|---|

Create and attach a new disk    Attach an existing disk

Figure 2.4: Setting up storage options

7. After defining disks, we get to the networking settings. Here, we need to define the **Virtual network** and **Subnet** options that the VM will use. These two options are mandatory. You can choose to assign the **Public IP** address to the VM (you can choose to disable the **Public IP** address, create a new one, or assign an existing IP address). The last part of the network settings relates to **NIC network security group**, where we need to choose whether we are going to use no network security group, a basic one, or an advanced one. There is also another option where we will define whether we will allow public ports. We can also configure **Accelerated networking** or **Load balancing** as additional options. An example of these VM network settings is shown in *Figure 2.5*:

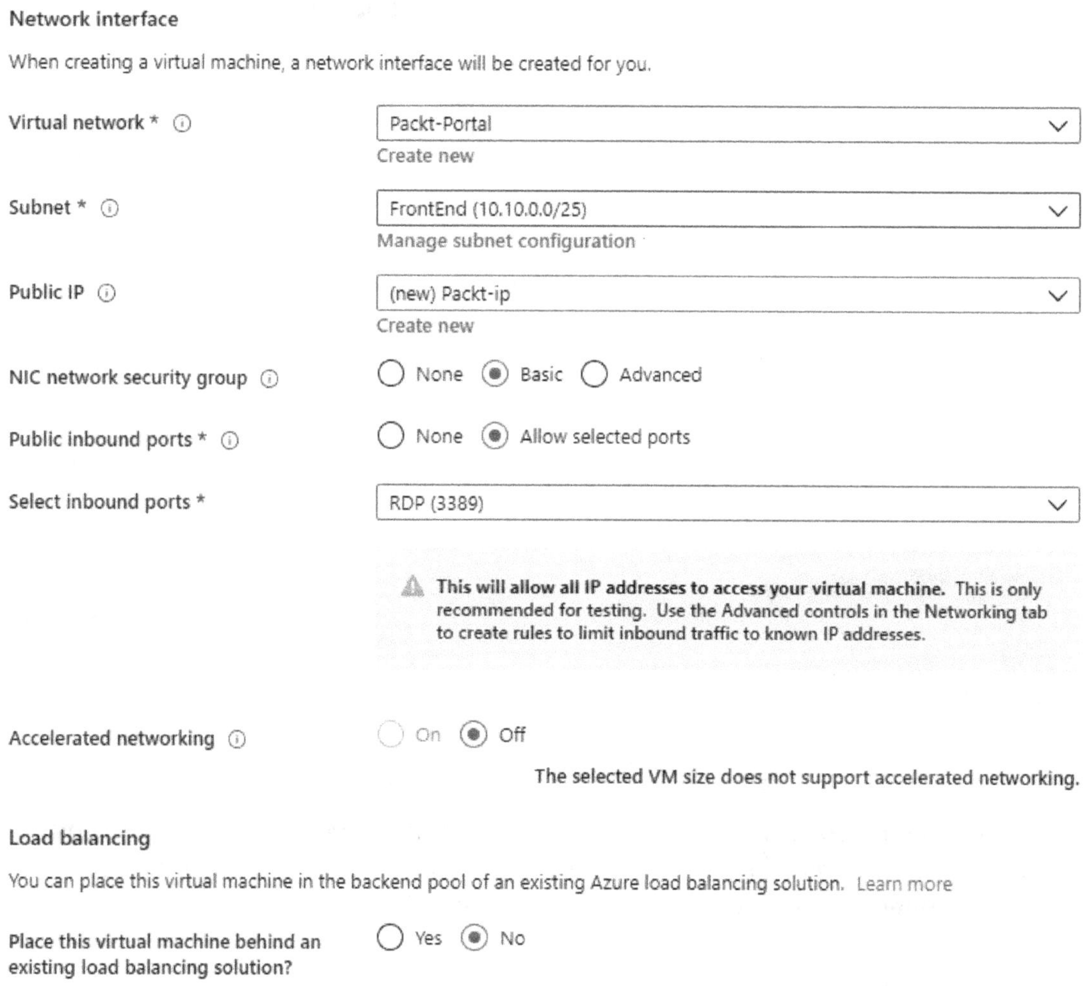

## Network interface

When creating a virtual machine, a network interface will be created for you.

Virtual network * ⓘ

Packt-Portal ⌄

Create new

Subnet * ⓘ

FrontEnd (10.10.0.0/25) ⌄

Manage subnet configuration

Public IP ⓘ

(new) Packt-ip ⌄

Create new

NIC network security group ⓘ  ◯ None  ◉ Basic  ◯ Advanced

Public inbound ports * ⓘ  ◯ None  ◉ Allow selected ports

Select inbound ports *

RDP (3389) ⌄

⚠ **This will allow all IP addresses to access your virtual machine.** This is only recommended for testing. Use the Advanced controls in the Networking tab to create rules to limit inbound traffic to known IP addresses.

Accelerated networking ⓘ  ◯ On  ◉ Off

The selected VM size does not support accelerated networking.

## Load balancing

You can place this virtual machine in the backend pool of an existing Azure load balancing solution. Learn more

Place this virtual machine behind an existing load balancing solution?  ◯ Yes  ◉ No

**Figure 2.5: Defining the virtual network and subnet options**

8. After the networking section, we need to set up **Management** as shown in *Figure 2.6*:

Monitoring

Boot diagnostics ⓘ      ◉ On   ○ Off

OS guest diagnostics ⓘ      ○ On   ◉ Off

Diagnostics storage account * ⓘ     [ (new) packtnetworkingportal468      ⌄ ]
Create new

Identity

System assigned managed identity ⓘ    ○ On   ◉ Off

Auto-shutdown

Enable auto-shutdown ⓘ      ○ On   ◉ Off

Backup

Enable backup ⓘ      ○ On   ◉ Off

**Figure 2.6: Enabling management features**

9. In **Advanced options**, we can set up post-deployment configuration steps by adding software installations, configuration scripts, custom data, and more. The **Advanced options** screen is shown in *Figure 2.7*:

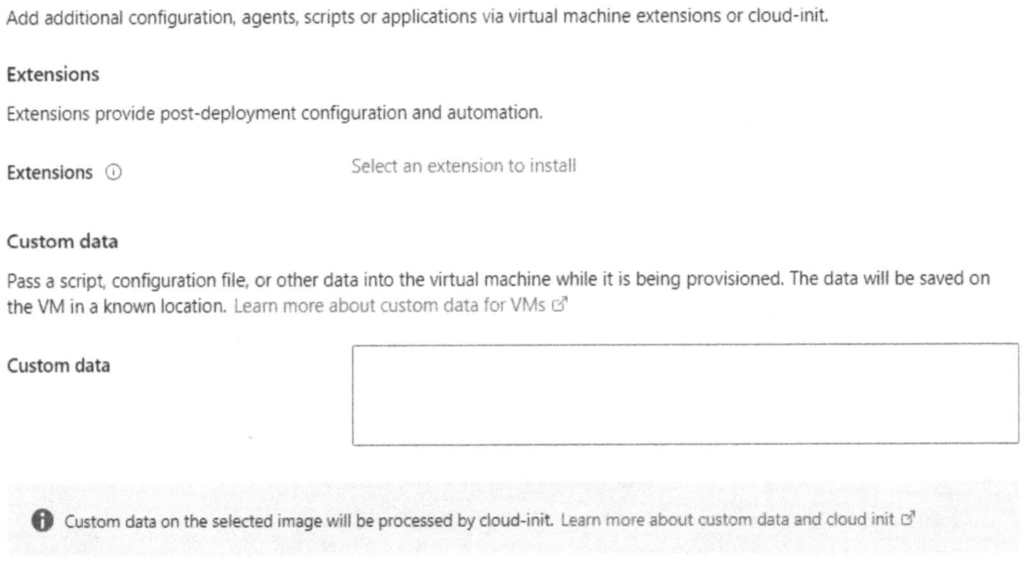

Add additional configuration, agents, scripts or applications via virtual machine extensions or cloud-init.

Extensions

Extensions provide post-deployment configuration and automation.

Extensions ⓘ       Select an extension to install

Custom data

Pass a script, configuration file, or other data into the virtual machine while it is being provisioned. The data will be saved on the VM in a known location. Learn more about custom data for VMs ↗

Custom data

ⓘ Custom data on the selected image will be processed by cloud-init. Learn more about custom data and cloud init ↗

**Figure 2.7: Setting up post-deployment configuration**

10. In the second part of **Advanced options**, we can select a **Host group** setting (this option provides a dedicated host that allows us to provision and manage a physical server in an Azure datacenter), a **Proximity placement group** (for grouping servers in the same region), and whether we want to use VMs from **Gen 1** or **Gen 2**. The default options are shown in *Figure 2.8*:

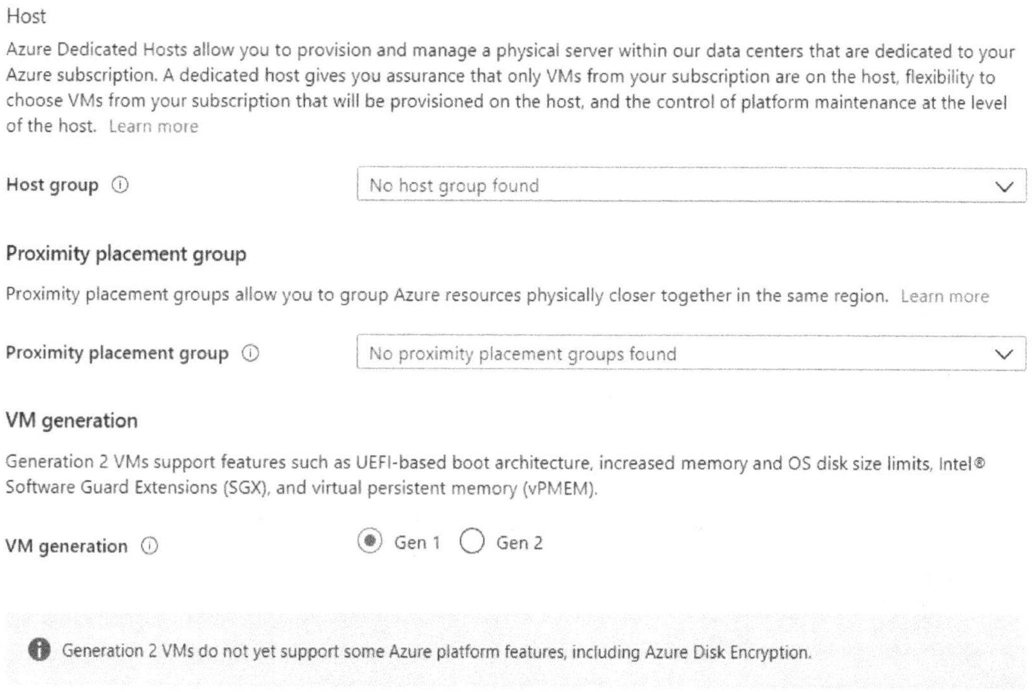

Host

Azure Dedicated Hosts allow you to provision and manage a physical server within our data centers that are dedicated to your Azure subscription. A dedicated host gives you assurance that only VMs from your subscription are on the host, flexibility to choose VMs from your subscription that will be provisioned on the host, and the control of platform maintenance at the level of the host.  Learn more

Host group  ⓘ                     | No host group found                                              ∨ |

Proximity placement group

Proximity placement groups allow you to group Azure resources physically closer together in the same region.  Learn more

Proximity placement group  ⓘ      | No proximity placement groups found                              ∨ |

VM generation

Generation 2 VMs support features such as UEFI-based boot architecture, increased memory and OS disk size limits, Intel® Software Guard Extensions (SGX), and virtual persistent memory (vPMEM).

VM generation  ⓘ               ⦿ Gen 1   ◯ Gen 2

ⓘ Generation 2 VMs do not yet support some Azure platform features, including Azure Disk Encryption.

**Figure 2.8: Allotting a dedicated host to provision and manage a physical server**

11. The last setting that we can edit concerns tags. Tags apply additional metadata to Azure resources to logically organize them into a taxonomy. The **Tags** tab is shown in *Figure 2.9*:

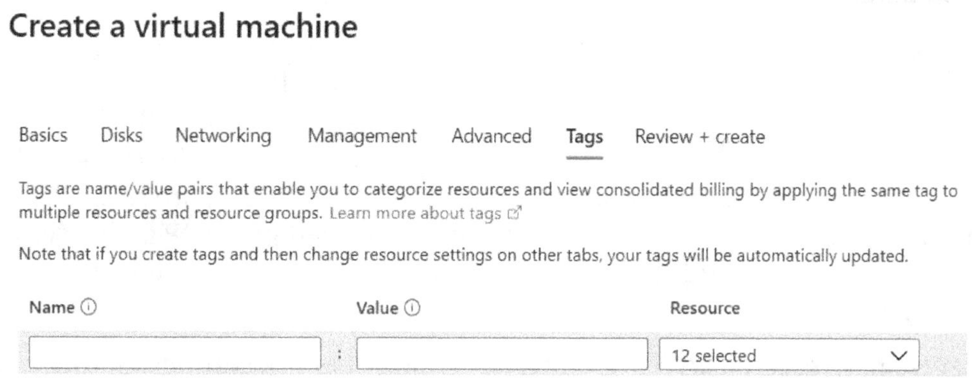

## Create a virtual machine

Basics   Disks   Networking   Management   Advanced   **Tags**   Review + create

Tags are name/value pairs that enable you to categorize resources and view consolidated billing by applying the same tag to multiple resources and resource groups. Learn more about tags ⧉

Note that if you create tags and then change resource settings on other tabs, your tags will be automatically updated.

Name ⓘ                          Value ⓘ                          Resource

|                             | : |                             | | 12 selected          ∨ |

**Figure 2.9: Applying tags to Azure resources**

12. After all the settings are defined, we get to the validation screen, where all our settings are checked for the last time. After validation is passed, we confirm the creation of a VM by clicking the **Create** button, as shown in *Figure 2.10*:

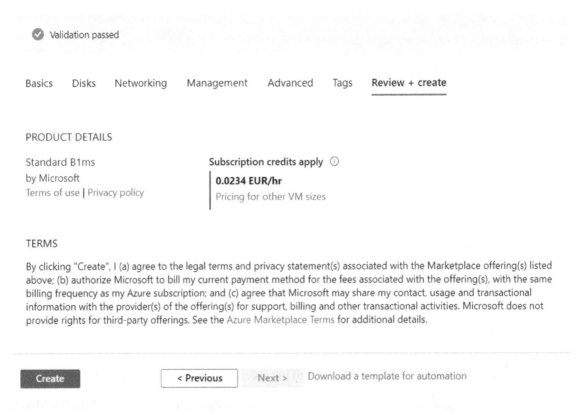

Figure 2.10: Creation of a VM

## How it works...

When a VM is created, an NIC is created in the process. An NIC is used as a sort of interconnection between the VM and the virtual network. An NIC is assigned a private IP address by the network. As an NIC is associated with both the VM and the virtual network, the IP address is used by the VM. Using this IP address, the VM can communicate over a private network with other VMs (or other Azure resources) on the same network. Additionally, NICs and VMs can be assigned public IP addresses as well. A public address can be used to communicate with the VM over the internet, either to access services or to manage the VM.

Now that we have created an Azure VM and defined network settings; in the next section, we'll see how to review these network settings.

## There's more...

If you are interested in finding out more about Azure VMs, you can read my book, Hands-On Cloud Administration in Azure, from Packt Publishing, where VMs are covered in more detail.

# Viewing VM network settings

After an Azure VM is created, we can review the network settings in the VM pane.

## Getting ready

Before you start, open a web browser and go to the Azure portal at https://portal.azure.com. Here, locate the previously created VM.

## How to do it...

In order to review the VM network settings, we must follow the steps given here:

1.  In the VM pane, locate the **Networking** settings. Here, you can see **Network interface**, **Application security groups**, and the **Network security group** associated with the VM. An example of this is shown in *Figure 2.11*:

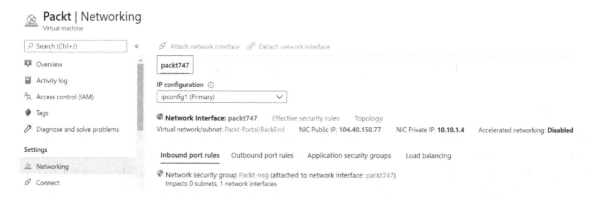

Figure 2.11: Network settings of a VM

2. If we select any of the associated network elements, we can discover more details. For example, if we select the **Network Interface** option associated with the VM, we can see other networking information such as **Private IP address**, **Public IP address**, **Virtual network/subnet**, **Network security group**, **IP configurations**, **DNS servers**, and more. The NIC view is shown in *Figure 2.12*:

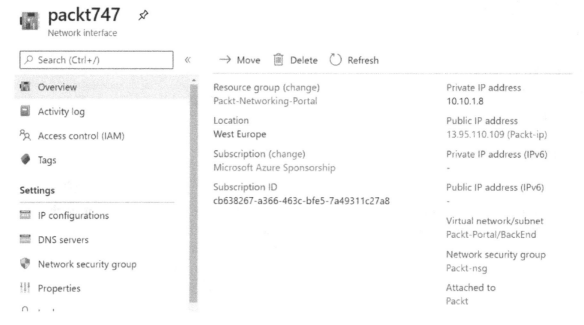

Figure 2.12: Viewing networking information from the NIC

## How it works...

Networking information is displayed in several places, including in the VM's network settings. Additionally, each Azure resource has a separate pane and exists as an individual resource, so we can view these settings in multiple places. However, the most complete picture of VM network settings can be found in the VM pane and the NIC pane.

# Creating a new NIC

An NIC is usually created during the VM creation process, but each VM can have multiple NICs. Based on this, we can create an NIC as an individual resource and attach it or detach it as needed.

## Getting ready

Before you start, open a web browser and go to the Azure portal at https://portal.azure.com.

## How to do it...

In order to create a new NIC using the Azure portal, we must take the following steps:

1. In the Azure portal, select **Create a resource** and choose **Network interface** under **Networking** services (or search for `network interface` in the search bar).

2. In the creation pane, we need to provide information for the **Name** and **Virtual network** fields, as well as giving the subnet that the NIC will be associated with. Other information to be provided includes the IP address assignment type (**Dynamic** or **Static**), whether we want the NIC to be associated with a **Network security group** type, and whether we want to use **IPv6**. All Azure resources require information on the **Subscription**, **Resource group**, and **Region**, and NICs are no exception. The information needed to create a new NIC is shown in *Figure 2.13*:

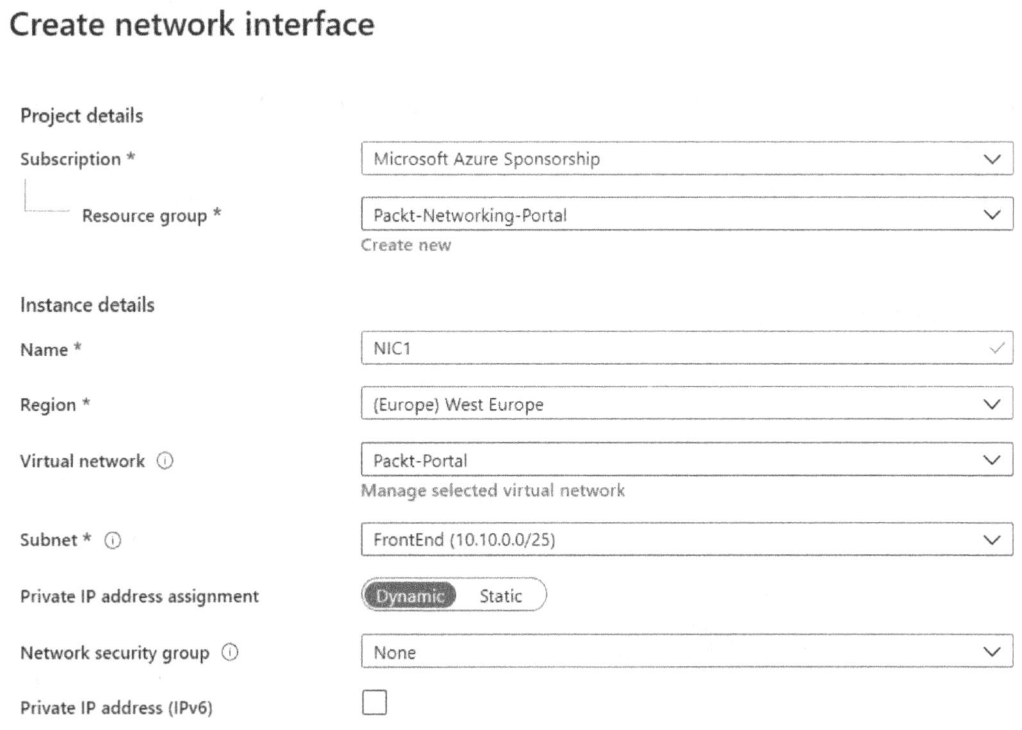

### Create network interface

**Project details**

| | |
|---|---|
| Subscription * | Microsoft Azure Sponsorship |
| Resource group * | Packt-Networking-Portal |
| | Create new |

**Instance details**

| | |
|---|---|
| Name * | NIC1 |
| Region * | (Europe) West Europe |
| Virtual network ⓘ | Packt-Portal |
| | Manage selected virtual network |
| Subnet * ⓘ | FrontEnd (10.10.0.0/25) |
| Private IP address assignment | **Dynamic**   Static |
| Network security group ⓘ | None |
| Private IP address (IPv6) | ☐ |

Figure 2.13: Creating an NIC using the Azure portal

## How it works...

An NIC can't exist without a network association, and this association must be assigned to a virtual network and subnet. This is defined during the creation process and cannot be changed later. On the other hand, association with a VM can be changed and the NIC can be attached or detached from a VM at any time.

# Attaching an NIC to a VM

Each VM can have multiple NICs. Because of this, we can add a new NIC at any time.

## Getting ready

Before you start, open a web browser and go to the Azure portal at https://portal.azure. com. Here, locate the VM we created earlier in this chapter.

## How to do it...

To attach an NIC to a VM, we must do the following:

1. In the VM pane, make sure the VM is stopped (that is, deallocated).

2. Locate the **Networking** settings in the VM pane.

3. At the top of the **Networking** settings screen in the VM pane, select the **Attach network interface** option.

4. A new option will appear, allowing you to create a new NIC or select an already-existing NIC that is not associated with the VM.

5. Click **OK** and, in a few moments, the process will finish and the NIC will be associated with the VM. An example of this is shown in *Figure 2.14*:

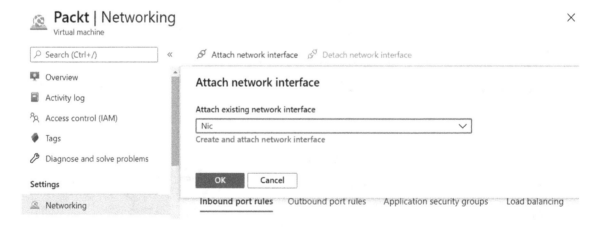

Figure 2.14: Attaching an NIC

## How it works...

Each VM can have multiple NICs. The number of NICs that can be associated with a VM depends on the type and size of the VM. To attach an NIC to a VM, the VM needs to be stopped (that is, deallocated); you can't add an additional NIC to a running VM.

# Detaching an NIC from a VM

Just as with attaching an NIC, we can detach an NIC at any time and attach it to another VM.

## Getting ready

Before you start, open a web browser and go to the Azure portal at https://portal.azure. com. Here, locate the previously created VM.

## How to do it...

To detach an NIC from a VM, we must do the following:

1.  In the VM pane, make sure the VM is stopped (that is, deallocated).

2.  Locate the **Networking** settings in the VM pane.

3.  At the top of the **Networking** settings screen in the VM pane, select the **Detach network interface** option.

4.  Select the NIC you want to detach from the VM.

5.  Click **OK** and, in a few moments, the process will finish and the NIC will be removed from the VM. An example of this is shown in *Figure 2.15*:

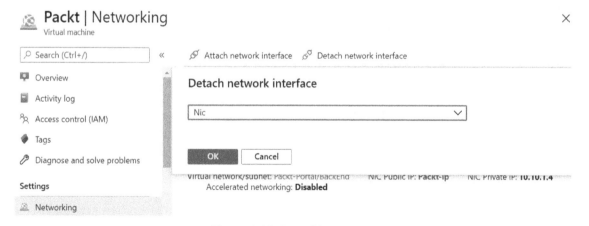

Figure 2.15: Detaching an NIC

## How it works...

To detach an NIC, the VM associated with the NIC must be stopped (that is, deallocated). At least one NIC must be associated with the VM—so you can't remove the last NIC from a VM. All network associations stay with the NIC—they are assigned to the NIC, not to the VM.

# 3

# Network Security Groups

**Network Security Groups (NSGs)** are built-in tools for network control that allow us to control incoming and outgoing traffic on a network interface or at the subnet level. They contain sets of rules that allow or deny specific traffic to specific resources or subnets in Azure. An NSG can be associated with either a subnet (by applying security rules to all resources associated with the subnet) or a **Network Interface Card (NIC)**, which is done by applying security rules to the **Virtual Machine (VM)** associated with the NIC.

We will cover the following recipes in this chapter:

- Creating a new NSG in the Azure portal
- Creating a new NSG with PowerShell
- Creating a new allow rule in an NSG
- Creating a new deny rule in an NSG
- Creating a new NSG rule with PowerShell
- Assigning an NSG to a subnet
- Assigning an NSG to a network interface
- Assigning an NSG to a subnet with PowerShell
- Creating an **Application Security Group (ASG)**
- Associating an ASG with a VM
- Creating rules with an NSG and an ASG

## Technical requirements

For this chapter, the following is required:

- An Azure subscription
- Azure PowerShell

The code samples can be found at https://github.com/PacktPublishing/Azure-Networking-Cookbook-Second-Edition/tree/master/Chapter03.

## Creating a new NSG in the Azure portal

As a first step to more effectively control network traffic, we are going to create a new NSG.

### Getting ready

Before you start, open your browser and go to the Azure portal, at https://portal.azure.com.

## How to do it...

To create a new NSG using the Azure portal, we must follow these steps:

1. In the Azure portal, select **Create a resource** and choose **Network security group** under **Networking** (or search for `network security group` in the search bar).

2. The parameters we need to define for the deployment are **Subscription**, **Resource group**, **Name**, and **Region**. An example of the required parameters is shown in *Figure 3.1*:

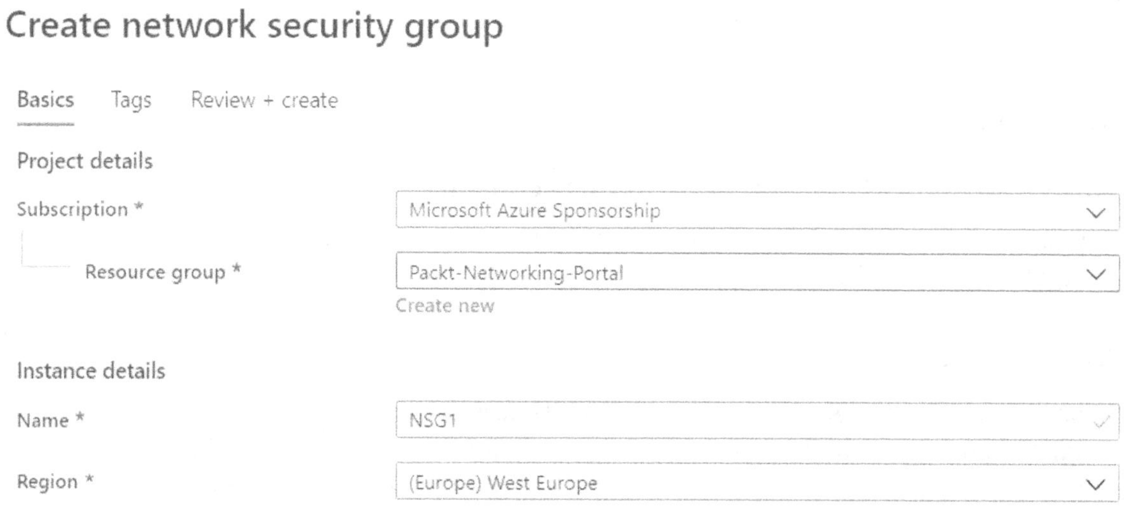

Create network security group

Basics    Tags    Review + create

Project details

Subscription *          Microsoft Azure Sponsorship

Resource group *        Packt-Networking-Portal
                        Create new

Instance details

Name *                  NSG1

Region *                (Europe) West Europe

Figure 3.1: Creating a new NSG using the Azure portal

After the deployment has been validated and started (it takes a few moments to complete), the NSG is ready for use.

## How it works...

The NSG deployment can be initiated during a VM deployment. This will associate the NSG to the NIC associated with the deployed VM. In this case, the NSG is already associated with the resource, and rules defined in the NSG will apply only to the associated VM.

If the NSG is deployed separately, as seen in this recipe, it is not associated and the rules that are created within it are not applied until an association has been created with the NIC or the subnet. When it is associated with a subnet, the NSG rules will apply to all resources on the subnet.

Let's move on to the next recipe to understand how to create a new NSG using PowerShell.

# Creating a new NSG with PowerShell

Alternatively, we can create an NSG using PowerShell. The advantage of this approach is that we can add NSG rules in a single script, creating custom rules right after the NSG is created. This allows us to automate the deployment process and create our own *default* rules right after the NSG has been created.

## Getting ready

Open the PowerShell console and make sure you are connected to your Azure subscription. Refer to *Chapter 1*, *Azure Virtual Network*, for a refresher on how to do this.

## How to do it...

To deploy a new NSG, execute the following command:

```
New-AzNetworkSecurityGroup -Name "nsg1" -ResourceGroupName "Packt-Networking-
Script" -Location "westeurope"
```

## How it works...

The script is using the **Resource Group** (**RG**) that was deployed in *Chapter 1*, *Azure Virtual Network* (we will use the same RG for all deployments). Otherwise, a new RG needs to be deployed prior to executing the script. The final outcome will be the same as creating a new NSG using the Azure portal: a new NSG will be created with default rules. An advantage of using PowerShell is that we can add additional rules during deployment that will help automate the process. You will see an example of this in the *Creating a new NSG rule with PowerShell* recipe later in this chapter.

In this recipe, you learned to create a new NSG using PowerShell. Let's move on to the next recipe to learn how to allow rules in NSG using the Azure portal.

# Creating a new allow rule in an NSG

When a new NSG is created, only the default rules are present, which allow all outbound traffic and block all inbound traffic. To change these, additional rules need to be created. First, we are going to show you how to create a new rule to allow inbound traffic.

## Getting ready

Before you start, open your browser and go to the Azure portal at https://portal.azure.com. Locate the previously created NSG.

## How to do it...

To create a new NSG allow rule using the Azure portal, we must follow these steps:

1.  In the **NSG** pane, locate the **Inbound security rules** option under **Settings**.

2.  Click on the **Add** button at the top of the page and wait for the new pane, to open:

Figure 3.2: Creating a new NSG allow rule using the Azure portal

3. In the new pane, we need to provide information for the **Source** (location and port range), **Destination** (location and port range), **Protocol**, **Action**, **Priority**, **Name**, and **Description** fields. If you want to allow traffic, make sure you select **Allow** for **Action**. An example of how to create a rule to allow traffic over port **443** (thus allowing traffic to the web server) is shown in *Figure 3.3*:

Figure 3.3: Creating a rule to allow traffic over port 443

## How it works...

By default, all traffic coming from Azure Load Balancer or Azure Virtual Network is allowed. All traffic coming over the internet is denied. To change this, we need to create additional rules. Make sure you set the right priority when creating rules. Rules with the highest priority (that is, those with the lower number) are processed first, so if you have two rules, one of which is denying traffic and one of which is allowing it, the rule with higher priority will take precedence, while the one with lower priority will be ignored.

In this recipe, you learned how to create a new rule to allow inbound traffic. In the next recipe, you will learn how to create a new rule in NSG to deny traffic.

# Creating a new deny rule in an NSG

When a new NSG is created, only the default rules are present. The default rules allow all outbound traffic and block all inbound traffic. To change this, additional rules need to be created. Now, we are going to show you how to create a new outbound rule to deny traffic.

## Getting ready

Before you start, open your browser and go to the Azure portal at https://portal.azure. com. Locate the previously created NSG.

## How to do it...

To create a new NSG deny rule using the Azure portal, we must follow these steps:

1. In the **NSG** pane, locate the **Outbound security rules** option under **Settings**.

2. Click on the **Add** button at the top of the page and wait for the new pane to open:

**Figure 3.4: Creating a new NSG deny rule using the Azure portal**

3. In the new pane, we need to provide information for **Source** (location and port range), **Destination** (location and port range), **Protocol**, **Action**, **Priority**, **Name**, and **Description**. If you want to deny traffic, make sure you select **Deny** for **Action**. An example of how to create a rule to deny traffic over port **22** is shown in *Figure 3.5*:

Figure 3.5: Adding an outbound security rule

## How it works...

All outbound traffic is allowed by default, regardless of where it is going. If we want to explicitly deny traffic on a specific port, we need to create a rule to do so. Make sure you set the priority right when creating rules. Rules with the highest priority (those with the lowest numbers) are processed first, so if you have two rules where one is denying traffic and one is allowing it, the rule with higher priority will apply.

Let's move on to the next recipe, where you will learn how to create an NSG rule using PowerShell.

# Creating a new NSG rule with PowerShell

Alternatively, we can create an NSG rule using PowerShell. This command can be executed right after the NSG has been created, allowing us to create and configure an NSG in a single script. This way, we can standardize deployment and have rules applied each time an NSG is created.

## Getting ready

Open the PowerShell console and make sure you are connected to your Azure subscription.

## How to do it...

To create a new NSG rule, execute the following command:

```
$nsg = Get-AzNetworkSecurityGroup -Name 'nsg1' -ResourceGroupName 'Packt-
Networking-Script'

$nsg | Add-AzNetworkSecurityRuleConfig -Name 'Allow_HTTPS' -Description
'Allow_HTTPS' -Access Allow -Protocol Tcp -Direction Inbound -Priority 100
-SourceAddressPrefix Internet -SourcePortRange * -DestinationAddressPrefix *
-DestinationPortRange 443 | Set-AzNetworkSecurityGroup
```

## How it works...

Using a script, creating an NSG rule is just a matter of parameters. The **Access** parameter, which can be either **Allow** or **Deny**, will determine whether we want to allow traffic or deny it. The **Direction** parameter, which can be **Inbound** or **Outbound**, determines whether the rule is for inbound or outbound traffic. All other parameters are the same, no matter what kind of rule we want to create. Again, priority plays a very important role, so we must make sure it's chosen correctly.

## There's more...

As mentioned in the *Creating a new NSG with PowerShell* recipe, we can create the NSG and the rules that are needed in a single script. The following script is an example of this:

```
$nsg = New-AzNetworkSecurityGroup -Name 'nsg1' -ResourceGroupName 'Packt-
Networking-Script' -Location "westeurope"

$nsg | Add-AzNetworkSecurityRuleConfig -Name 'Allow_HTTPS' -Description
'Allow_HTTPS' -Access Allow -Protocol Tcp -Direction Inbound -Priority 100
-SourceAddressPrefix Internet -SourcePortRange * -DestinationAddressPrefix *
-DestinationPortRange 443 | Set-AzNetworkSecurityGroup
```

This recipe explained how to create a new NSG rule using PowerShell. In the next recipe, you will learn how to assign an NSG to a subnet.

# Assigning an NSG to a subnet

The NSG and its rules must be assigned to a resource to have any impact. Here, you are going to see how to associate an NSG with a subnet.

## Getting ready

Before you start, open your browser and go to the Azure portal at https://portal.azure. com. Locate the previously created NSG.

## How to do it...

To assign an NSG to a subnet, follow these steps:

1. In the NSG pane, locate the **Subnets** option under **Settings**.

2. Click on the **Associate** button at the top of the page and wait for the new pane to open:

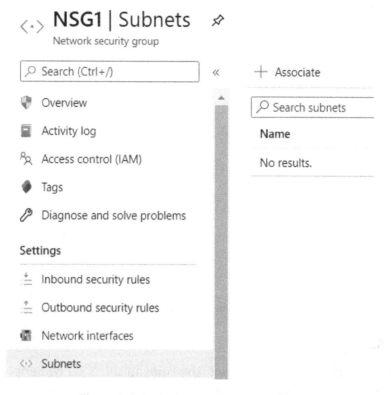

Figure 3.6: Assigning an NSG to a subnet

3. In the new pane, first select the virtual network that contains the subnet you want to associate the NSG with, and then select the subnet, as seen in *Figure 3.7*:

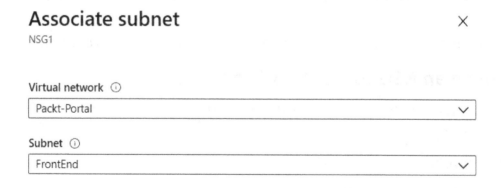

Figure 3.7: Associating the subset with the NSG

4.  After submitting the change, the subnet will appear in a list of associated subnets:

Figure 3.8: A list of associated subnets

## How it works...

When an NSG is associated with a subnet, the rules in the NSG will apply to all of the resources in the subnet. Note that the subnet can be associated with more than one NSG, and the rules from all the NSGs will apply in that case. Priority is the most important factor when looking at a single NSG, but when the rules from more NSGs are observed, the **Deny** rule will prevail. So, if we have two NSGs on a subnet, one with **Allow** on port **443** and another one with the **Deny** rule on the same port, traffic on this port will be denied.

Let's move on to the next recipe and learn how to assign an NSG to a network interface.

# Assigning an NSG to a network interface

Now, we are going to widen our scope and show you how to associate an NSG with a network interface.

## Getting ready

Before you start, open your browser and go to the Azure portal at https://portal.azure.com. Locate the previously created NSG.

## How to do it...

To assign an NSG to a network interface, follow these steps:

1.  In the NSG pane, locate the **Network interfaces** option under **Settings**.

2. Click on the **Associate** button at the top of the page and wait for the new pane to open:

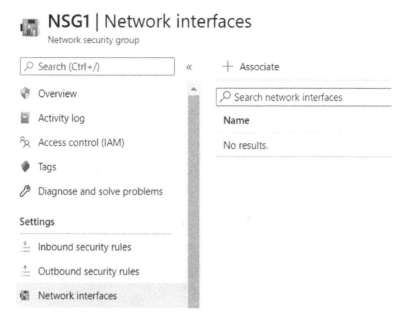

Figure 3.9: Assigning the NSG to a network interface

3. Select the NIC you want to associate the NSG with from the list of those available:

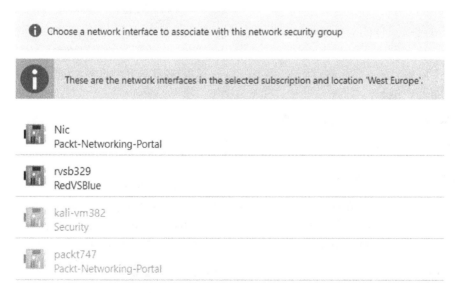

Figure 3.10: Associating with the network interface

## How it works...

When an NSG is associated with an NIC, the NSG rules will apply only to a single NIC (or a VM associated with the NIC). The NIC can be associated with only one NSG directly, but a subnet associated with an NIC can have an association with another NSG (or even multiple NSGs). This is similar to when we have multiple NSGs assigned to a single subnet, and the **Deny** rule will take higher priority. If one of the NSGs allows traffic on a port, but another NSG is blocking it, traffic will be denied.

In this recipe, you learned how to assign an NSG to a network interface. Let's move on to the next recipe, where you will learn how to assign an NSG using PowerShell.

# Assigning an NSG to a subnet with PowerShell

Alternatively, we can associate an NSG using Azure PowerShell. In this recipe, we are going to show you how to associate an NSG with a subnet.

## Getting ready

Open the PowerShell console and make sure you are connected to your Azure subscription.

## How to do it...

To associate an NSG with a subnet, execute the following command:

```
$vnet = Get-AzVirtualNetwork -Name 'Packt-Script' -ResourceGroupName 'Packt-
Networking-Script'

$subnet = Get-AzVirtualNetworkSubnetConfig -VirtualNetwork $vnet -Name
BackEnd

$nsg = Get-AzNetworkSecurityGroup -ResourceGroupName 'Packt-Networking-
Script' -Name 'nsg1'

$subnet.NetworkSecurityGroup = $nsg

Set-AzVirtualNetwork -VirtualNetwork $vnet
```

## How it works...

To assign an NSG using PowerShell, we need to collect information on the virtual network, subnet, and NSG. When all of the information is gathered, we can perform the association using the **Set-AzVirtualNetwork** command and apply the changes.

Let's move on to the next recipe and create an ASG using the Azure portal.

# Creating an Application Security Group (ASG)

ASGs are an extension of NSGs, allowing us to create additional rules and take better control of traffic. Using only NSGs allows us to create rules that will allow or deny traffic only for a specific source, IP address, or subnet. ASGs allow us to create better filtering and create additional checks on what traffic is allowed based on ASGs. For example, with NSGs, we can create a rule that subnet A can communicate with subnet B. If we have the application structure for it and an associated ASG, we can add resources in application groups. By adding this element, we can create a rule that will allow communication between subnet A and subnet B, but only if the resources belong to the same application.

## Getting ready

Before you start, open your browser and go to the Azure portal at https://portal.azure.com.

## How to do it...

To create an ASG using the Azure portal, we must follow these steps:

1. In the Azure portal, select **Create a resource** and choose **Application security group** under **Networking** (or search for `application security group` in the search bar).

2. The parameters we need to define for deployment are **Subscription**, **Resource group**, **Name**, and **Region**. An example of the required parameters is shown in *Figure 3.11*:

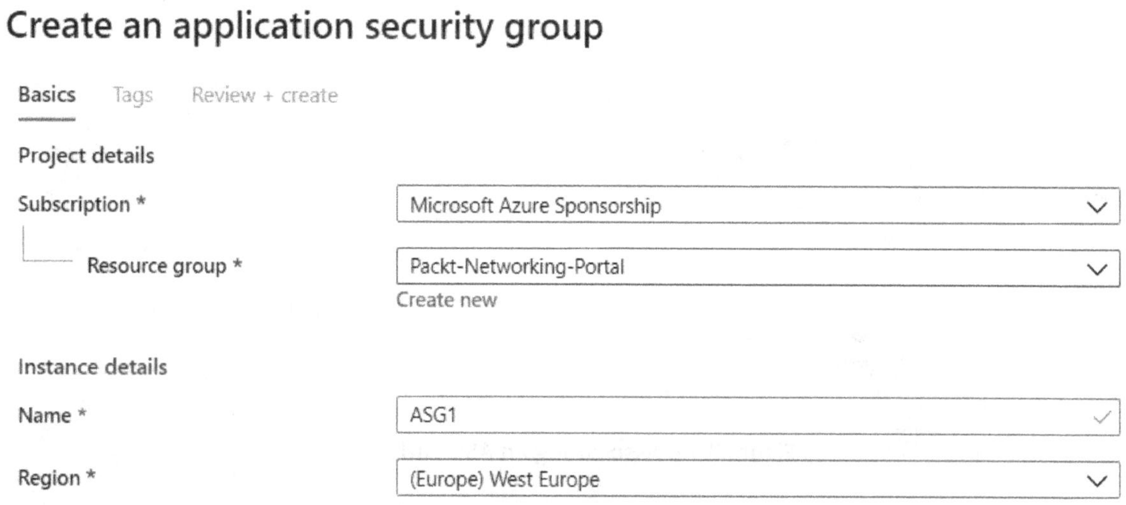

## Create an application security group

Basics    Tags    Review + create

**Project details**

| | |
|---|---|
| Subscription * | Microsoft Azure Sponsorship |
| Resource group * | Packt-Networking-Portal |
| | Create new |

**Instance details**

| | |
|---|---|
| Name * | ASG1 |
| Region * | (Europe) West Europe |

Figure 3.11: Creating an ASG using the Azure portal

## How it works...

ASGs don't make much difference on their own and must be combined with NSGs to create NSG rules that will allow better control of traffic, applying additional checks before traffic flow is allowed.

Now that we have created an ASG, let's move on to a new recipe where we will associate the ASG with a VM.

# Associating an ASG with a VM

After creating an ASG, we must associate it with a VM. After this is done, we can create rules with the NSG and ASG for traffic control.

## Getting ready

Before you start, open your browser and go to the Azure portal at https://portal.azure. com. Locate the previously created VM.

## How to do it...

To associate an ASG with a VM, we must follow these steps:

1.  In the VM pane, locate the **Networking** settings.

2.  In the **Networking** settings, select the **Application security groups** tab, as shown in *Figure 3.12*:

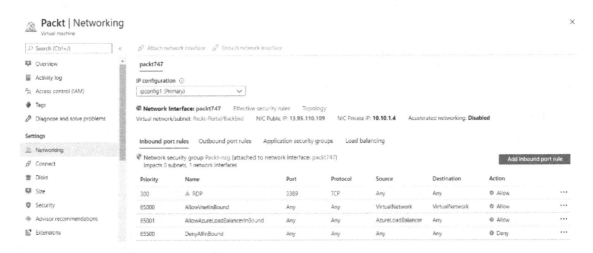

Figure 3.12: Associating an ASG with a VM

3. In the **Application security groups** settings, select **Configure the application security groups**, as shown in *Figure 3.13*:

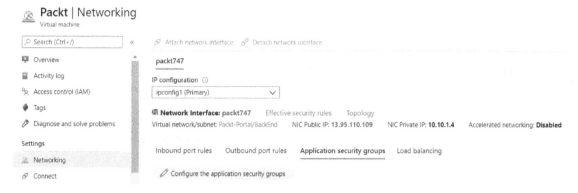

Figure 3.13: Configuring ASGs

4. In the new pane from the list of available ASGs, select the ASG that you want to associate the VM with:

Figure 3.14: Associating an ASG with a VM

5. After clicking **Save**, it takes a few seconds to apply the changes, after which the VM will be associated with the ASG.

## How it works...

The VM must be associated with the ASG. We can associate more than one VM with each ASG. The ASG is then used in combination with the NSG to create new NSG rules.

In the next recipe, we will create new rules using an NSG and an ASG.

# Creating rules with an NSG and an ASG

As a final step, we can use NSGs and ASGs to create new rules with better control. This approach allows us to have better control of traffic, limiting incoming traffic not only to a specific subnet but also only based on whether or not the resource is part of the ASG.

## Getting ready

Before you start, open your browser and go to the Azure portal at https://portal.azure.com. Locate the previously created NSG.

## How to do it...

To create a rule using both an ASG and an NSG, we must follow these steps:

1.  In the NSG pane, find **Inbound security rules**. Select **Add** to add a new rule.

2.  For the source, select **Application Security Group**, and then select the ASG you want to use as the source. We also need to provide parameters for **Source**, **Source port ranges**, **Destination**, **Destination port ranges**, **Protocol**, **Action**, **Priority**, **Name**, and **Description**. An example is shown in *Figure 3.15*:

**Figure 3.15: Adding an inbound security rule**

## How it works...

Using only NSGs to create rules, we can allow or deny traffic only for a specific IP address or range. With an ASG, we can widen or narrow this as needed. For example, we can create a rule to allow VMs from a frontend subnet, but only if these VMs are in a specific ASG. Alternatively, we can allow access to a number of VMs from different virtual networks and subnets, but only if they belong to a specific ASG.

# 4

# Managing IP addresses

In Azure, we can have two types of IP addresses, private and public. Public addresses can be accessed over the internet. Private addresses are from the Azure Virtual Network address space and are used for private communication on private networks. Addresses can be assigned to a resource or can exist as a separate resource.

We will cover the following recipes in this chapter:

- Creating a new public IP address in the Azure portal
- Creating a new public IP address with PowerShell
- Assigning a public IP address
- Unassigning a public IP address
- Creating a reservation for a public IP address
- Removing a reservation for a public IP address
- Creating a reservation for a private IP address
- Changing a reservation for a private IP address
- Removing a reservation for a private IP address
- Adding multiple addresses to an NIC
- Creating a public IP prefix

## Technical requirements

For this chapter, the following is required:

- An Azure subscription
- Azure PowerShell

The code samples can be found at https://github.com/PacktPublishing/Azure-Networking-Cookbook-Second-Edition/tree/master/Chapter04.

## Creating a new public IP address in the Azure portal

Public IP addresses can be created as a separate resource or created during the creation of some other resources (a **virtual machine** (**VM**), for example). Therefore, a public IP can exist as part of a resource or as a standalone resource. First, we are going to show you how to create a new public IP address.

### Getting ready

Before you start, open your browser and go to the Azure portal at https://portal.azure.com.

## How to do it...

To create a new public IP address, we must follow these steps:

1. In the Azure portal, select **Create a resource** and choose **Public IP address** under **Networking** services (or search for `public IP address` in the search bar).

2. The parameters we need to define for deployment are **IP Version**, **SKU**, **Name**, **IP address assignment**, **DNS name label**, **Subscription**, **Resource group**, and **Location**. Idle timeout (the amount of time that the connection will stay open with no activity) is defaulted to 4 minutes but can be increased to 30 minutes at most. An example of the required parameters is shown in *Figure 4.1*:

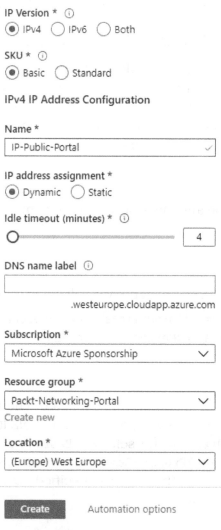

**Figure 4.1: Creating a new public IP address using the Azure portal**

## How it works...

The **stock keeping unit (SKU)** can be either **Basic** or **Standard**. The main differences are that **Standard** is closed to inbound traffic by default (inbound traffic must be whitelisted in **Network Security Groups (NSGs)**) and that **Standard** is zone redundant. Another difference is that a **Standard** SKU public IP address has a static assignment, while a **Basic** SKU can be either static or dynamic.

You can choose either the **IPv4** or **IPv6** version for the IP address, or both, but choosing **IPv6** will limit you to a dynamic assignment for the **Basic** SKU and static assignment for the **Standard** SKU.

The **DNS name label** is optional—it can be used to resolve the endpoint if dynamic assignment is selected. Otherwise, there is no point in creating a DNS label, as an IP address can always be used to resolve the endpoint if static assignment is selected.

# Creating a new public IP address with PowerShell

Alternatively, we can create a public IP address using Azure PowerShell. Again, this approach is best when we want to automate the process. Even though a public IP address can exist on its own, it's usually created to be associated with other resources and to be used as an endpoint. When using PowerShell to create a resource, we can continue to the next step and join it with a resource in a single script.

## Getting ready

Open the PowerShell console and make sure you are connected to your Azure subscription.

## How to do it...

To deploy a new public IP address, execute the following command:

```
New-AzPublicIpAddress -Name 'ip-public-script' -ResourceGroupName 'Packt-
Networking-Script' -AllocationMethod Dynamic -Location 'westeurope'
```

## How it works...

As an outcome, a new public IP address will be created. The settings, in this case, will be a basic SKU dynamic assignment, IPv4 version, and no DNS label. Furthermore, we can use additional switches such as **-SKU** for selecting **Basic** or **Standard**, **-IPAddressVersion** for choosing between **IPv4** and **IPv6**, or **-DomainNamelabel** to specify the DNS label. These are optional parameters—if these aren't specified, Azure will create the public IP with the aforementioned default values.

# Assigning a public IP address

A public IP address can be created as a separate resource or disassociated from another resource and exist on its own. Such an IP address can then be assigned to a new resource or another already-existing resource. If the resource is no longer in use or has been migrated, we can still use the same public IP address. In this case, the public endpoint that's used to access a service may stay unchanged. This can be useful when a publicly available application or service is migrated or upgraded, as we can keep using the same endpoint and users don't need to be aware of any change.

## Getting ready

Before you start, open your browser and go to the Azure portal at https://portal.azure. com.

## How to do it...

To assign a public IP address, we must do the following:

1. Locate the **network interface (NIC)** that you want the IP address to be assigned to. This can be done directly by finding the NIC, or through the VM pane that the NIC is assigned to.

2. In the **Network interface** pane, go to **IP configurations** under **Settings**, and select the configuration shown in *Figure 4.2*:

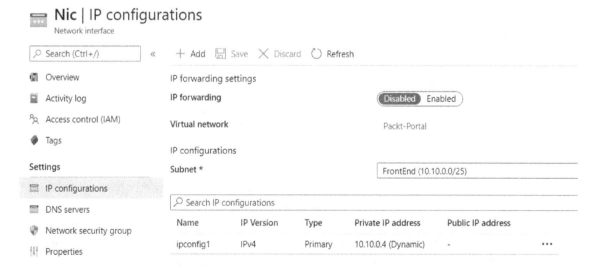

Figure 4.2: Viewing the IP configurations in the NIC pane

3. In the new pane, select **Associate** under **Public IP address** and select the **Public IP address** that you want to assign from the drop-down menu. Only unassigned IP addresses in the same region will show in the list. An example of this is shown in *Figure 4.3*:

Figure 4.3: Assigning a public IP address

4. After the public IP address has been selected, click **Save** to apply the settings.

## How it works...

A public IP address exists as a separate resource and can be assigned to a resource at any time. When a public IP address is assigned, you can use this IP address to access services running on a resource that the IP address is assigned to (remember that an appropriate NSG must be applied). We can also remove an IP address from a resource and assign it to a new resource. For example, if we want to migrate services to a new VM, the IP address can be removed from the old VM and assigned to the new one. This way, service endpoints running on the VM will not change. This is especially useful when static IP addresses are used.

# Unassigning a public IP address

A public IP address can be unassigned from a resource in order to be saved for later use or assigned to another resource. When a resource is deleted or decommissioned, we can still put the public IP address to use and assign it to the next resource.

## Getting ready

Before you start, open your browser and go to the Azure portal at https://portal.azure. com. Make sure that the VM using a public IP address is not running.

## How to do it...

To unassign a public IP address, we must do the following:

1. Locate the NIC that the public IP address is associated with.

2. In the **Network interface** pane, go to **IP configurations** under **Settings** and select the **IP configuration**:

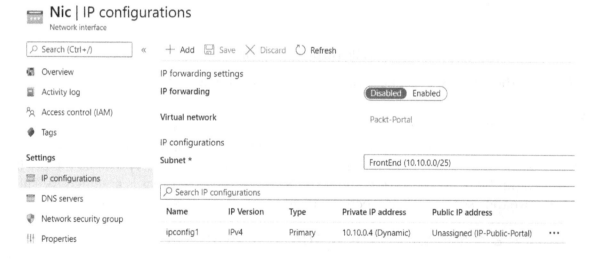

Figure 4.4: IP configurations under the NIC pane

3. In the new pane, change the **Public IP address** setting to **Disassociate**:

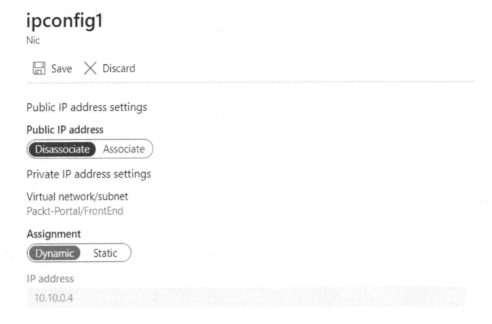

Figure 4.5: Unassigning the public IP address

4. After the changes are made, click **Save** to apply the new configuration.

## How it works...

A public IP address can be assigned or unassigned from a resource in order to save it for future use or to transfer it to a new resource. To remove it, we simply disable the public IP address in the IP configuration under the NIC that the IP address is assigned to. This will remove the association but keep the IP address as a separate resource.

# Creating a reservation for a public IP address

The default option for a public IP address is dynamic IP assignment. This can be changed during the public IP address creation, or later. If this is changed from dynamic IP assignment, then the public IP address becomes reserved (or static).

## Getting ready

Before you start, open your browser and go to the Azure portal at https://portal.azure.com.

## How to do it...

To create a reservation for a public IP address, follow these steps:

1. Locate the public IP address in the Azure portal. This can be done by finding the IP address directly, or through the resource it's assigned to (either the NIC or VM).

2. In the **Public IP address** pane, go to **Configuration** under **Settings**. Change **Assignment** from **Dynamic** to **Static**, as shown in *Figure 4.6*:

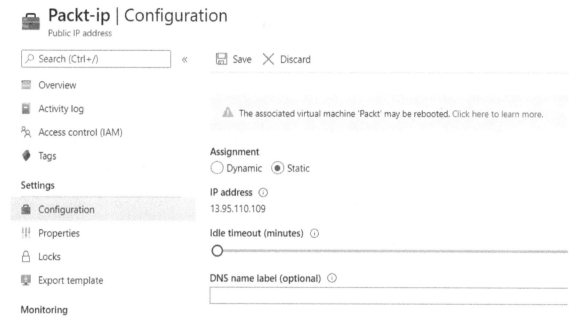

Figure 4.6: Changing the public IP address assignment to Static

3. After this change has been made, click **Save** to apply the new settings.

## How it works...

A public IP address is set to dynamic by default. This means that an IP address might change in time. For example, if a VM that an IP address is assigned to is turned off or rebooted, there is a possibility that the IP address will change after the VM is up and running again. This can cause issues if services that are running on the VM are accessed over the public IP address, or if there is a DNS record associated with the public IP address.

We create an IP reservation and set the assignment to static to avoid such a scenario and keep the IP address reserved for our services.

# Removing a reservation for a public IP address

If the public IP address is set to static, we can remove a reservation and set the IP address assignment to dynamic. This isn't done often as there is usually a reason why the reservation is set in the first place. But as the reservation for the public IP address has an additional cost, there is sometimes a need to remove the reservation if it is not necessary.

## Getting ready

Before you start, open your browser and go to the Azure portal at https://portal.azure. com. Make sure that the IP address is not associated with any resource.

## How to do it...

To remove a reservation for a public IP address, follow these steps:

1. Locate the public IP address in the Azure portal.

2. In the **Public IP address** pane, go to **Configuration** under **Settings** and set **Assignment** to **Dynamic**:

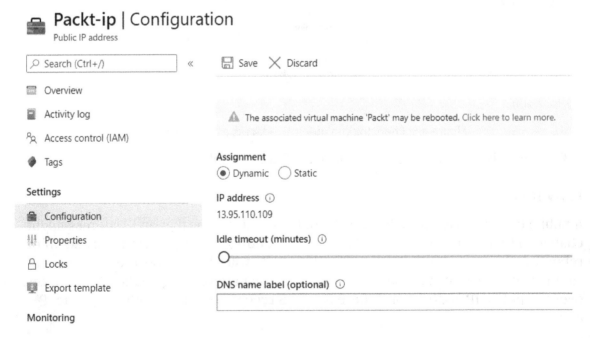

Figure 4.7: Changing the public IP address assignment to Dynamic

3. After these changes have been made, click **Save** to apply the new configuration.

## How it works...

To remove an IP reservation from a public IP address, the public IP address must not be associated with a resource. We can remove the reservation by setting the IP address assignment to dynamic.

The main reason for this is pricing. In Azure, the first five public IP reservations are free. After the initial five, each new reservation is billed. To avoid paying anything unnecessary, we can remove a reservation when it is not needed or when the public IP address is not being used.

# Creating a reservation for a private IP address

Similar to public IP addresses, we can make a reservation for private IP addresses. This is usually done to ensure communication between servers on the same virtual network and to allow the usage of IP addresses in connection strings.

## Getting ready

Before you start, open your browser and go to the Azure portal at https://portal.azure.com.

## How to do it...

To create a reservation for a private IP address, follow these steps:

1. In the Azure portal, locate the NIC you want to make the reservation for.

2. In the **Network interface** pane, go to **IP configurations** under **Settings** and select the IP configuration:

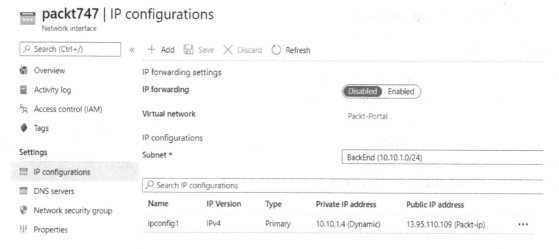

Figure 4.8: Viewing IP configurations in the NIC pane

3.  In the new pane, under the **Private IP address** settings, set **Assignment** to **Static**. The current IP address value will be set automatically. If needed, you can change that value to another value, but it must be in the address space of the subnet associated with the NIC:

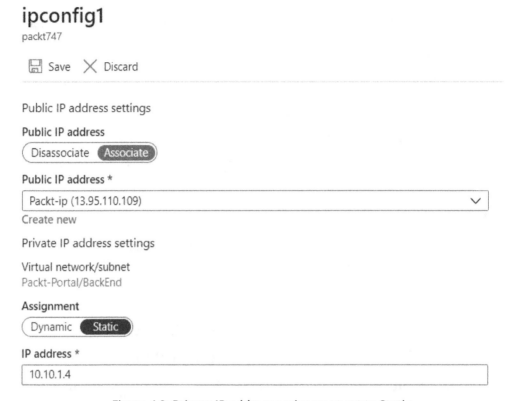

Figure 4.9: Private IP address assignment set to Static

4.  After these changes have been made, click **Save** to apply the new configuration.

## How it works...

A reservation can be made for private IP addresses. The difference is that a private IP address does not exist as a separate resource but is assigned to an NIC.

Another difference is that you can select a value for a private IP address. A public IP address is assigned randomly and can be reserved, but you cannot choose which value this will be. For private IP addresses, you can select the value for the IP, but it must be an unused IP from the subnet associated with the NIC.

# Changing a reservation for a private IP address

For private IP addresses, you can change the IP address at any time to another value. With public IP addresses, this isn't the case, as you get the IP address randomly from a pool and aren't able to change the value. With a private IP address, you can change the value to another IP address from the address space.

## Getting ready

Before you start, open your browser and go to the Azure portal at https://portal.azure. com.

## How to do it...

To change a reservation for a private IP address, follow these steps:

1.  In the Azure portal, locate the NIC you want to make changes for.

2.  In the **Network interface** pane, go to **IP configurations** under **Settings** and select the IP configuration:

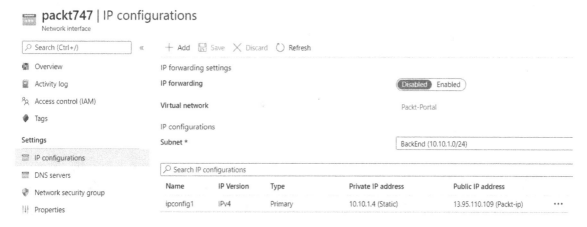

**Figure 4.10: Locating the IP configuration in the Network interface pane**

3. In **Private IP address settings**, enter a new value for **IP address**:

# ipconfig1

packt747

💾 Save    ✕ Discard

⚠️ The virtual machine associated with this network interface will be restarted to utilize the new private IP address. The network interface will be re-provisioned and network configuration settings, including secondary IP addresses, subnet masks, and default gateway, will need to be manually reconfigured within the virtual machine. Learn more

Public IP address settings

**Public IP address**

( Disassociate    Associate )

**Public IP address** *

Packt-ip (13.95.110.109)    ⌄

Create new

Private IP address settings

**Virtual network/subnet**
Packt-Portal/BackEnd

**Assignment**

( Dynamic    Static )

**IP address** *

10.10.1.8    ✓

Figure 4.11: Assigning a new value for the private IP address

4. After these changes have been made, click **Save** to apply the new configuration.

## How it works...

A reservation for a private IP address can be changed. Again, the value must be an unused IP address from a subnet associated with the NIC. If the VM associated with the NIC is turned off, the new IP address will be assigned upon its next startup. If the VM is running, it will be restarted to apply the new changes.

# Removing a reservation for a private IP address

Similar to public IP addresses, we can remove a reservation for a private IP address at any time. A private IP address is free, so additional costs aren't a factor in this case. But there are scenarios where dynamic assignment is required, and we can set it at any time.

## Getting ready

Before you start, open your browser and go to the Azure portal at https://portal.azure.com.

## How to do it...

To remove a reservation for a private IP address, follow these steps:

1. In the Azure portal, locate the NIC you want to make changes for.

2. In the **Network interface** pane, go to **IP configurations** under **Settings** and select the IP configuration:

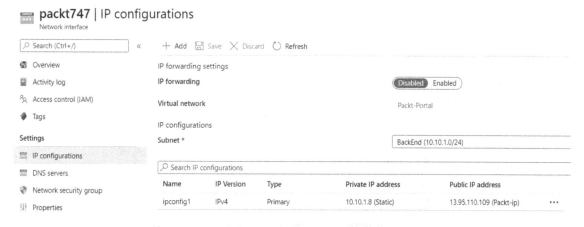

Figure 4.12: Selecting the IP configuration in the Network interface pane

3. In the new pane, under **Private IP address settings**, change **Assignment** to **Dynamic**:

Figure 4.13: Private IP address assignment set to Dynamic

4. After these changes have been made, click **Save** to apply the new configuration.

## How it works...

We can remove a private IP address reservation at any time by switching **Assignment** to **Dynamic**. When this change is made, the VM associated with the NIC will be restarted to apply the new changes. After a change is made, a private IP address may change after the VM is restarted or turned off.

# Adding multiple IP addresses to an NIC

In various situations, we may need to have multiple IP addresses associated with a single NIC. In Azure, this is possible for both private and public IP addresses.

## Getting ready

Before you start, open your browser and go to the Azure portal at https://portal.azure.com.

## How to do it...

1.  In the Azure portal, locate the NIC you want to make changes for.

2.  In the **Network interface** pane, go to **IP configurations** under **Settings** and click **Add**:

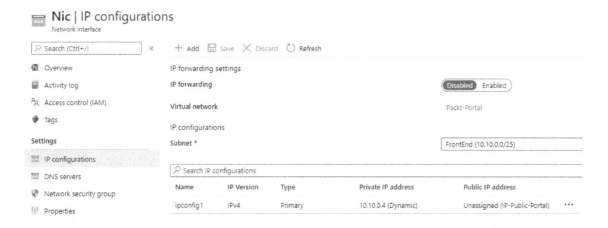

Figure 4.14: The Network interface pane

3. A new pane for IP configuration will appear. We need to provide values for the **Name** and **Type** fields (**Type** will be grayed out if another IP configuration already exists), and we need to select some IP address settings. If only a private IP address is needed, we just need to select the private address **Allocation** and click **Create**:

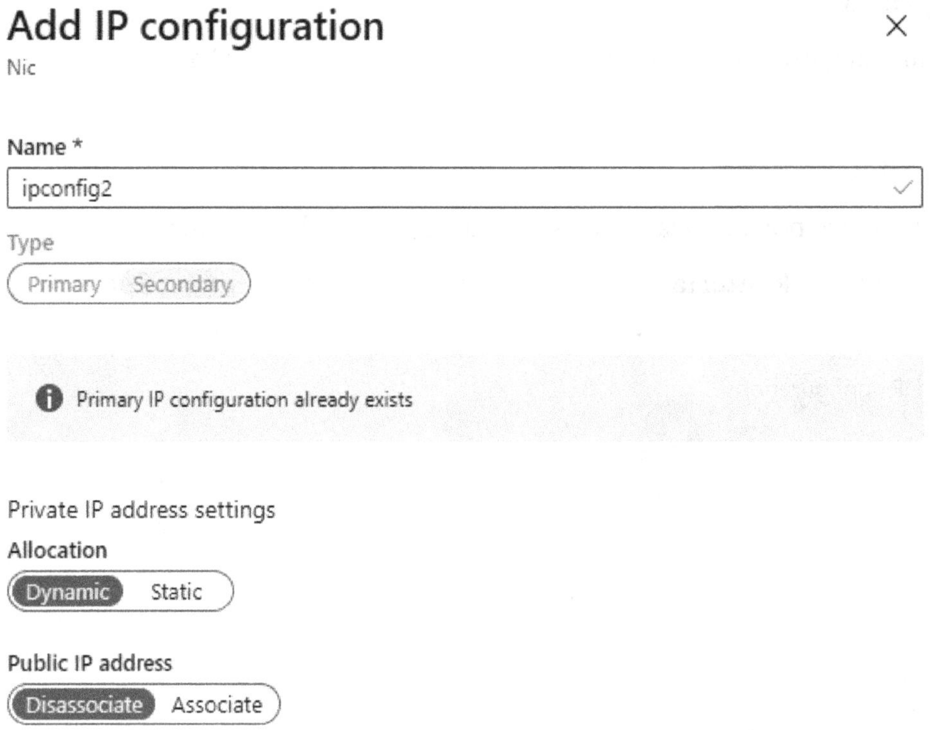

## Add IP configuration

Nic

Name *

> ipconfig2

Type

( Primary    Secondary )

ⓘ  Primary IP configuration already exists

Private IP address settings

Allocation

( Dynamic    Static )

Public IP address

( Disassociate    Associate )

Figure 4.15: Adding IP configuration to the NIC

4.  If an additional public IP address is needed, we need to select **Associate** under **Public IP address**. We are required to provide additional information for **Name**, **SKU**, and the **Assignment** type:

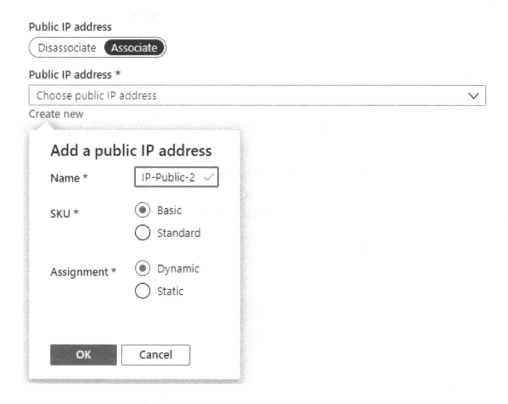

Figure 4.16: Adding a new public IP address

## How it works...

Each NIC can have multiple IP configurations assigned. Each IP configuration must have a private IP address and can have a public IP address. So, it is possible to add a private IP address without a public IP address, but not the other way around. This provides us with different routing options and the ability to communicate with different applications and services over different IP addresses. Routing will be explained in more detail in *Chapter 6: DNS and routing*.

# Creating a public IP prefix

Creating new resources is usually associated with creating new IP addresses. There can be issues when public IP addresses need to be associated with firewall rules or app configurations. To overcome this, we can create a public IP prefix and reserve a range of IP addresses that will be assigned to our resources.

## How to do it...

To create a new public IP prefix, we must follow these steps:

1. In the Azure portal, select **Create a resource** and choose **Public IP prefix** under **Networking** services (or search for `public IP prefix` in the search bar).

2. We need to provide information for **Subscription**, **Resource group**, **Name**, **Region**, and **IP Version**. **SKU** is not selectable and is set to **Standard**. For **Prefix size**, we define how many IP addresses we want to reserve:

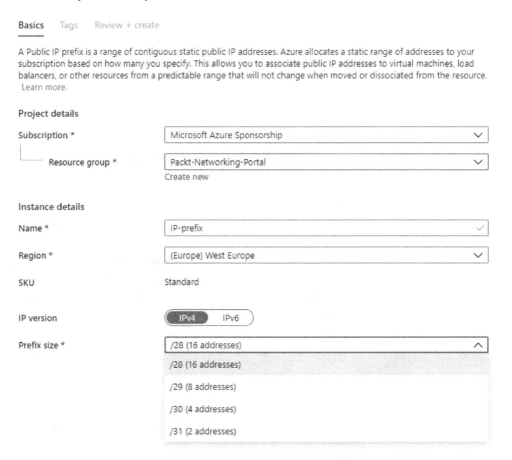

Figure 4.17: Creating a public IP prefix

## How it works...

When we create a public IP prefix, public IP address association is not done randomly but from a pool of addresses reserved for us. In many ways, this acts similarly to creating a virtual network and defining a private IP address space, only with public IP addresses. This can be very useful when we need to know addresses in advance. For example, let's say we need to create a firewall rule for each service we create. That would require us to wait for each service to be deployed and get a public IP address after it has been created. With a public IP prefix, IP addresses are known in advance and we can set a rule for a range of IP addresses, rather than doing it IP by IP.

# 5

# Local and virtual network gateways

Local and virtual network gateways are **virtual private network (VPN)** gateways that are used to connect to on-premises networks and encrypt all traffic going between an **Azure Virtual Network (VNet)** and a local network. Each virtual network can have only one virtual network gateway, but one virtual network gateway can be used to configure multiple VPN connections.

We will cover the following recipes in this chapter:

- Creating a local network gateway in the Azure portal
- Creating a local network gateway with PowerShell
- Creating a virtual network gateway in the Azure portal
- Creating a virtual network gateway with PowerShell
- Modifying the local network gateway settings

## Technical requirements

For this chapter, the following is required:

- An Azure subscription
- Azure PowerShell

The code samples can be found in https://github.com/PacktPublishing/Azure-Networking-Cookbook-Second-Edition/tree/master/Chapter05.

# Creating a local network gateway in the Azure portal

When a Site-to-Site connection is created, we have to provide configuration for both sides of the connection—that is, both Azure and on-premises. Although a local network gateway is created in Azure, it represents your local (on-premises) network and holds configuration information on your local network settings. It's an essential component for creating the VPN connection that is needed to create a Site-to-Site connection between the virtual network and the local network.

## Getting ready

Before you start, open a web browser and go to the Azure portal at https://portal.azure.com.

## How to do it...

In order to create a new local network gateway, the following steps are required:

1. In the Azure portal, select **Create a resource** and choose **Local network gateway** under the **Networking** services (or search for `local network gateway` in the search bar).

2. The parameters that we need to provide are **Name**, **IP address** (that is, the public IP address of the local firewall), **Address space** (the local address space that you want to connect to), **Subscription**, **Resource group**, and **Location**. Optionally, we can configure the **Border Gateway Protocol** (**BGP**) settings:

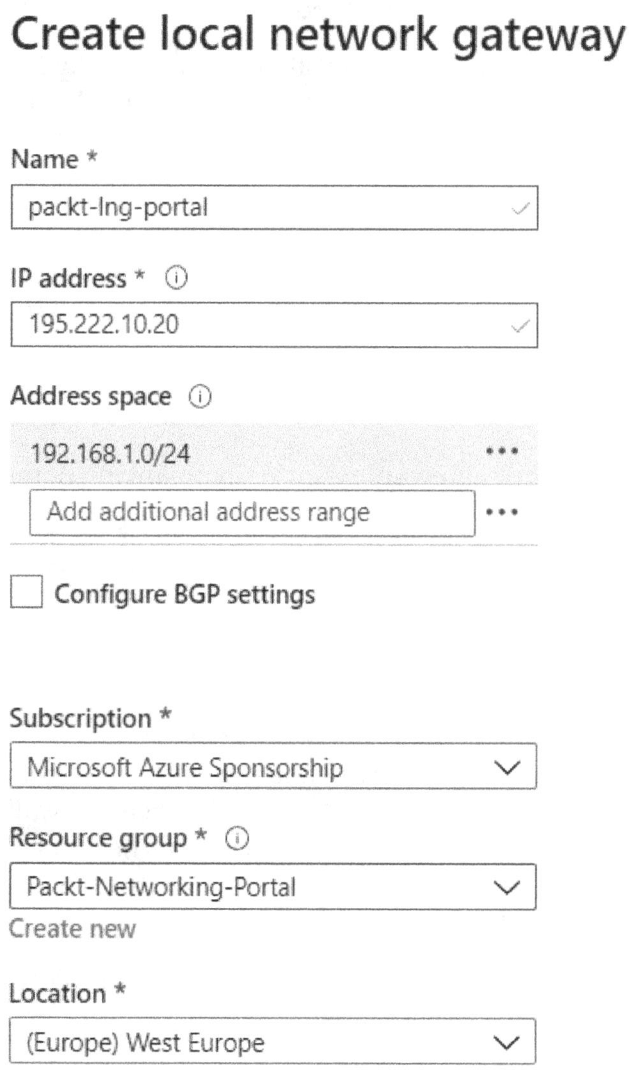

# Create local network gateway

Name *

packt-lng-portal

IP address *  ⓘ

195.222.10.20

Address space  ⓘ

192.168.1.0/24                          •••

Add additional address range           •••

☐ Configure BGP settings

Subscription *

Microsoft Azure Sponsorship            ⌄

Resource group *  ⓘ

Packt-Networking-Portal                ⌄
Create new

Location *

(Europe) West Europe                   ⌄

**Figure 5.1: Creating a new local network gateway**

## How it works...

The local network gateway is used to connect a virtual network gateway to an on-premises network. The virtual network gateway is directly connected to the virtual network and has all the relevant Azure VNet information needed to create a VPN connection. On the other hand, a local network gateway holds all the local network information needed to create a VPN connection.

In this recipe, we created a local network gateway in the Azure portal. In the next recipe, we will learn how to do the same using PowerShell.

# Creating a local network gateway with PowerShell

As mentioned in the previous recipe, the local network gateway holds information on the local network that we want to connect to an Azure VNet. In addition to creating a local network gateway through the Azure portal, we can create it with Azure PowerShell.

## Getting ready

Open the PowerShell console and make sure you are connected to your Azure subscription.

## How to do it...

To create a new local network gateway, execute the following command:

```
New-AzLocalNetworkGateway -Name packt-lng-script -ResourceGroupName 'Packt-
Networking-Script' -Location 'westeurope' -GatewayIpAddress '195.222.10.20'
-AddressPrefix '192.168.1.0/24'
```

## How it works...

In order to deploy a new local network gateway, we need to provide parameters for the name, resource group, location, gateway IP address, and address prefix that we want. The gateway IP address is the public IP address of the local firewall that you are trying to connect to. The address prefix is the subnet prefix of the local network that you are trying to connect to. This address must be associated with a firewall address that is provided as a gateway IP address.

In this recipe, we created a local network gateway with Azure PowerShell. Let's move on to the next recipe and learn how to create a virtual network gateway in the Azure portal.

# Creating a virtual network gateway in the Azure portal

After a local network gateway is created, we need to create a virtual network gateway in order to create a VPN connection between the local and Azure networks. As a local network gateway holds information on the local network, the virtual network gateway holds information for the Azure VNet that we are trying to connect to.

## Getting ready

Before you start, open a web browser and go to the Azure portal at https://portal.azure.com.

## How to do it...

In order to create a new virtual network gateway, the following steps are required:

1.  In the Azure portal, select **Create a resource** and choose **Virtual network gateway** under the **Networking** services (or search for `virtual network gateway` in the search bar).

2.  Everything is done in a single pane, but for the purpose of better visibility, I'm going to break it down into two sections. In the first section, we need to provide **Subscription**, **Name**, **Region**, **Gateway type**, **VPN type**, **SKU**, and **Generation**, (the **Generation** option depends on the SKU; not all SKUs support **Generation 2**), and we need to select **Virtual network** that will be used in the connection. Note that the gateway subnet must be created prior to this, and only virtual networks with a gateway subnet will be available for selection. An example is shown in *Figure 5.2*:

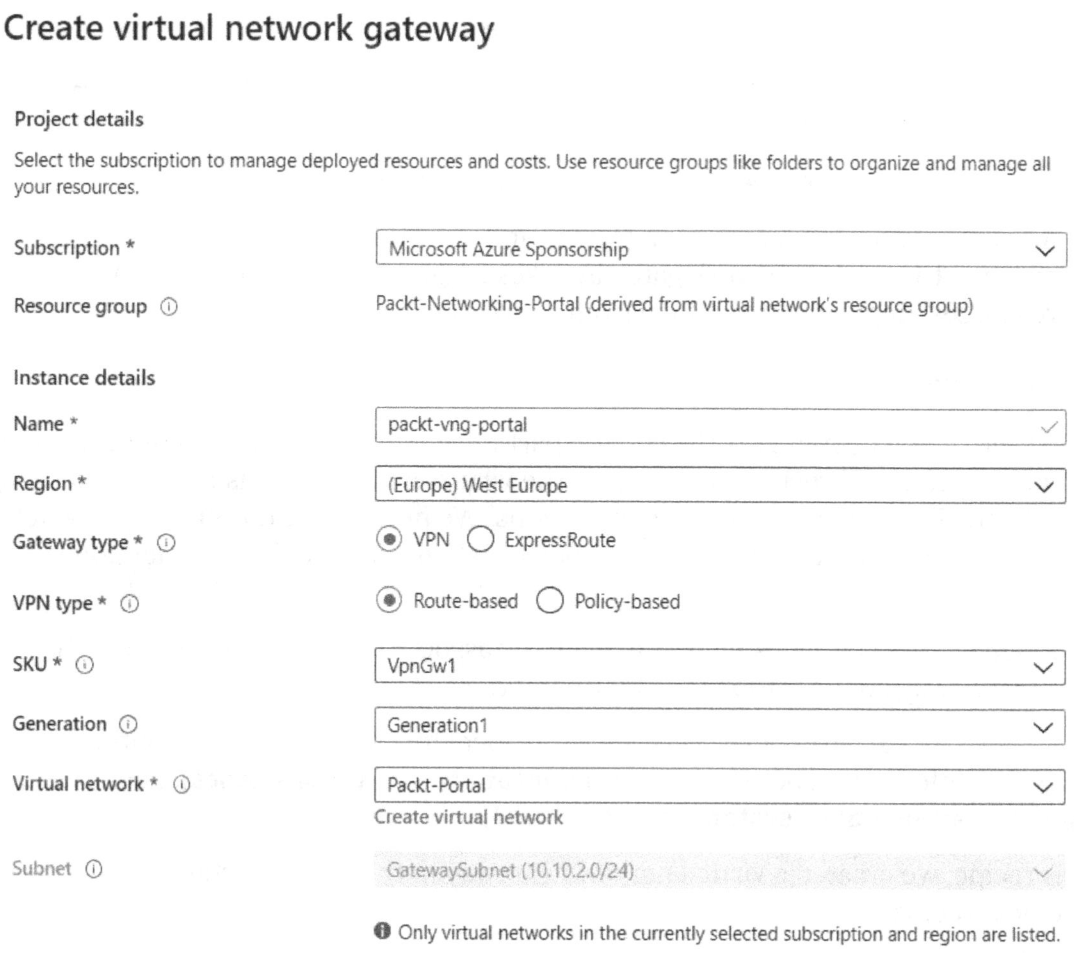

## Create virtual network gateway

**Project details**

Select the subscription to manage deployed resources and costs. Use resource groups like folders to organize and manage all your resources.

| | |
|---|---|
| Subscription * | Microsoft Azure Sponsorship |
| Resource group ⓘ | Packt-Networking-Portal (derived from virtual network's resource group) |

**Instance details**

| | |
|---|---|
| Name * | packt-vng-portal |
| Region * | (Europe) West Europe |
| Gateway type * ⓘ | ◉ VPN ◯ ExpressRoute |
| VPN type * ⓘ | ◉ Route-based ◯ Policy-based |
| SKU * ⓘ | VpnGw1 |
| Generation ⓘ | Generation1 |
| Virtual network * ⓘ | Packt-Portal |
| | Create virtual network |
| Subnet ⓘ | GatewaySubnet (10.10.2.0/24) |

ⓘ Only virtual networks in the currently selected subscription and region are listed.

**Figure 5.2: Creating a new virtual network gateway**

3.  In the second section, we need to set the **Public IP address** options (select an existing IP address or create a new one), and optionally, we can set **Enable active-active mode** and **Border Gateway Protocol Autonomous System Number (BGP ASN)**:

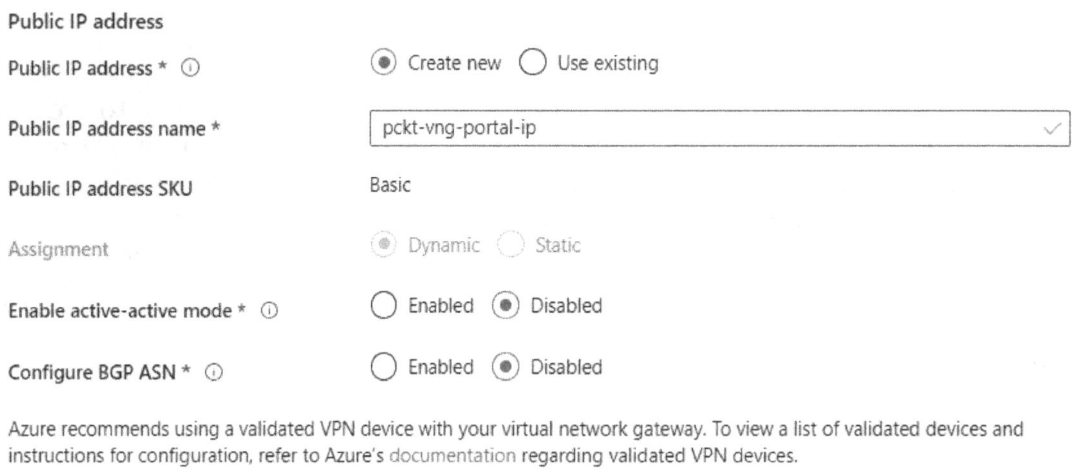

Public IP address

| | |
|---|---|
| Public IP address * ⓘ | ◉ Create new  ◯ Use existing |
| Public IP address name * | pckt-vng-portal-ip ✓ |
| Public IP address SKU | Basic |
| Assignment | ◉ Dynamic  ◯ Static |
| Enable active-active mode * ⓘ | ◯ Enabled  ◉ Disabled |
| Configure BGP ASN * ⓘ | ◯ Enabled  ◉ Disabled |

Azure recommends using a validated VPN device with your virtual network gateway. To view a list of validated devices and instructions for configuration, refer to Azure's documentation regarding validated VPN devices.

Figure 5.3: Setting the public IP address options

4.  After validation, we can click on **Create** and start the deployment. Note that creating the virtual network gateway takes longer than for most other Azure resources; deployment can take from 45 to 90 minutes.

## How it works...

The virtual network gateway is the second part needed to establish the connection to the Azure VNet. It is directly connected to the virtual network and is needed to create both Site-to-Site and Point-to-Site connections. We need to set the VPN type, which needs to match the type of the local VPN device when a Site-to-Site connection is created.

Active-active mode provides high availability by associating two IP addresses with separate gateway configurations to ensure uptime.

The border gateway protocol is a standard protocol for the exchange of routing and reachability information between different **autonomous systems (ASes)**. Each system is assigned an **autonomous systems number (ASN)**.

In this recipe, we created a virtual network gateway in the Azure portal. Let's move on to the next recipe.

# Creating a virtual network gateway with PowerShell

Creating a virtual network gateway is possible with PowerShell. Again, this helps automate processes. For example, if we start creating a virtual network gateway using a portal and notice that our virtual network isn't listed, it's probably because it's missing a gateway subnet. So, we must abandon the process, go back, create the gateway subnet, and start creating the virtual network gateway. Using PowerShell, we can ensure that all the requisite resources are present before starting, and then continue with the creation of the virtual network gateway.

## Getting ready

Open the PowerShell console and make sure that you are connected to your Azure subscription.

## How to do it...

To create a new virtual network gateway, execute the following script:

```
$vnet = Get-AzVirtualNetwork -ResourceGroupName 'Packt-Networking-Script'
-Name 'Packt-Script'

Add-AzVirtualNetworkSubnetConfig -Name 'GatewaySubnet' -AddressPrefix
10.11.2.0/27 -VirtualNetwork $vnet

$vnet | Set-AzVirtualNetwork

$gwpip = New-AzPublicIpAddress -Name VNet1GWIP -ResourceGroupName 'Packt-
Networking-Script' -Location 'westeurope' -AllocationMethod Dynamic


$vnet = Get-AzVirtualNetwork -ResourceGroupName 'Packt-Networking-Script'
-Name 'Packt-Script'

$subnet = Get-AzVirtualNetworkSubnetConfig -Name 'GatewaySubnet'
-VirtualNetwork $vnet

$gwipconfig = New-AzVirtualNetworkGatewayIpConfig -Name gwipconfig1 -SubnetId
$subnet.Id -PublicIpAddressId $gwpip.Id

New-AzVirtualNetworkGateway -Name VNet1GW -ResourceGroupName 'Packt-
Networking-Script' -Location 'westeurope' -IpConfigurations $gwipconfig
-GatewayType Vpn -VpnType RouteBased -GatewaySku VpnGw1
```

## How it works...

The script performs a few different operations to make sure that all requirements are met so that we can create a virtual network gateway. The first step is to collect information on the virtual network that we are going to use. Next, we add the gateway subnet to Azure VNet and create a public IP address that will be used by the virtual network gateway. We collect all the information and ensure that all the required resources are present, and then finally we create a new virtual network gateway.

In this recipe, we learned how to create a virtual network gateway with Azure PowerShell. In the next recipe, we will learn how to modify the settings of the local network gateway.

# Modifying the local network gateway settings

Network configurations may change over time, and we may need to address these changes in Azure as well—for example, the public IP address of a local firewall may change, and we'd then need to reconfigure the local network gateway, or a local network might be reconfigured and the address space or subnet has changed, so we would need to reconfigure the local network gateway once again.

## Getting ready

Before you start, open a web browser and go to the Azure portal at https://portal.azure.com.

## How to do it...

In order to modify local network gateway settings, we must do the following:

1. Locate the local network gateway in the Azure portal and go to **Configuration**.

2. In **Configuration**, we can edit **IP address** or **Address space**. We can also add additional address spaces if we want to connect multiple local subnets to Azure VNet:

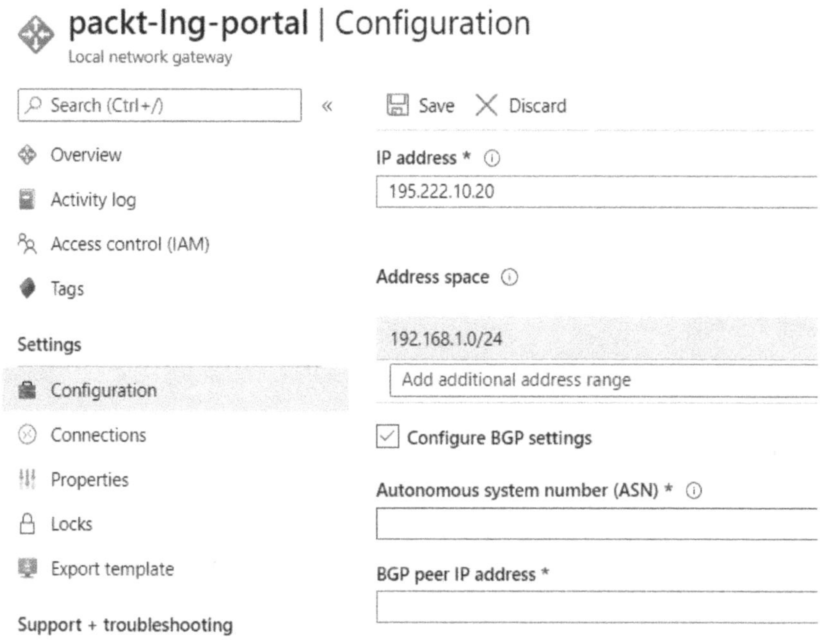

**packt-lng-portal** | Configuration
Local network gateway

Search (Ctrl+/) »

💾 Save ✕ Discard

- ⬥ Overview
- 📄 Activity log
- 👥 Access control (IAM)
- 🔖 Tags

**Settings**

- 🖥 Configuration
- ⊗ Connections
- ⫴ Properties
- 🔒 Locks
- 🖥 Export template

**Support + troubleshooting**

IP address * ⓘ

195.222.10.20

Address space ⓘ

192.168.1.0/24

Add additional address range

☑ Configure BGP settings

Autonomous system number (ASN) * ⓘ

BGP peer IP address *

**Figure 5.4: Modifying the local network gateway settings**

## How it works...

The local network gateway holds the local network information needed to create a Site-to-Site connection between the local and Azure networks. If this information changes, we can edit it in the **Configuration** settings. The changes that can be made are the IP address (that is, the public IP address of the local firewall) and the address space we are connecting to. Additionally, we can add or remove address spaces if we want to add or remove subnets that are able to connect to Azure VNet. If the configuration in the local network gateway is no longer valid, we can still use it to create a completely new connection to a new local network if needed.

# 6

# DNS and routing

Azure DNS allows us to host **Domain Name System (DNS)** domains in Azure. When using Azure DNS, we use Microsoft infrastructure for the name resolution, which results in fast and reliable DNS queries. Azure DNS infrastructure uses a vast number of servers to provide great reliability and availability of service. Using Anycast networking, each DNS query is answered by the closest available DNS server to provide a quick reply.

We will cover the following recipes in this chapter:

- Creating an Azure DNS zone
- Creating an Azure Private DNS zone
- Integrating a virtual network with a private DNS zone
- Creating a new record set in Azure DNS
- Creating a route table
- Changing a route table
- Associating a route table with a subnet
- Dissociating a route table from a subnet
- Creating a new route
- Changing a route
- Deleting a route

# Technical requirements

For this chapter, the following is required:

- An Azure subscription

# Creating an Azure DNS zone

To start using Azure DNS, we must first create a DNS zone. A DNS zone holds a DNS record for a specific domain, and it can hold records for a single domain at a time. A DNS zone will hold DNS records for this domain and possible subdomains. DNS name servers are set up to reply to any query on a registered domain and point to a destination.

## Getting ready

Before you start, open your browser and go to the Azure portal via https://portal.azure. com.

## How to do it...

In order to create a new Azure DNS zone with the Azure portal, we must follow these steps:

1. In the Azure portal, select **Create a resource** and choose **DNS Zone** under **Networking** services (or search for DNS Zone in the search bar).

2. In the new pane, we must enter information for the **Subscription**, **Resource group**, and **Name** fields. If we select an existing resource group, the region will automatically be the same as the one for the resource group selected. Optionally, we can mark this zone if the child of an existing zone is hosted in Azure DNS. The name must be a **Fully Qualified Domain Name (FQDN)**:

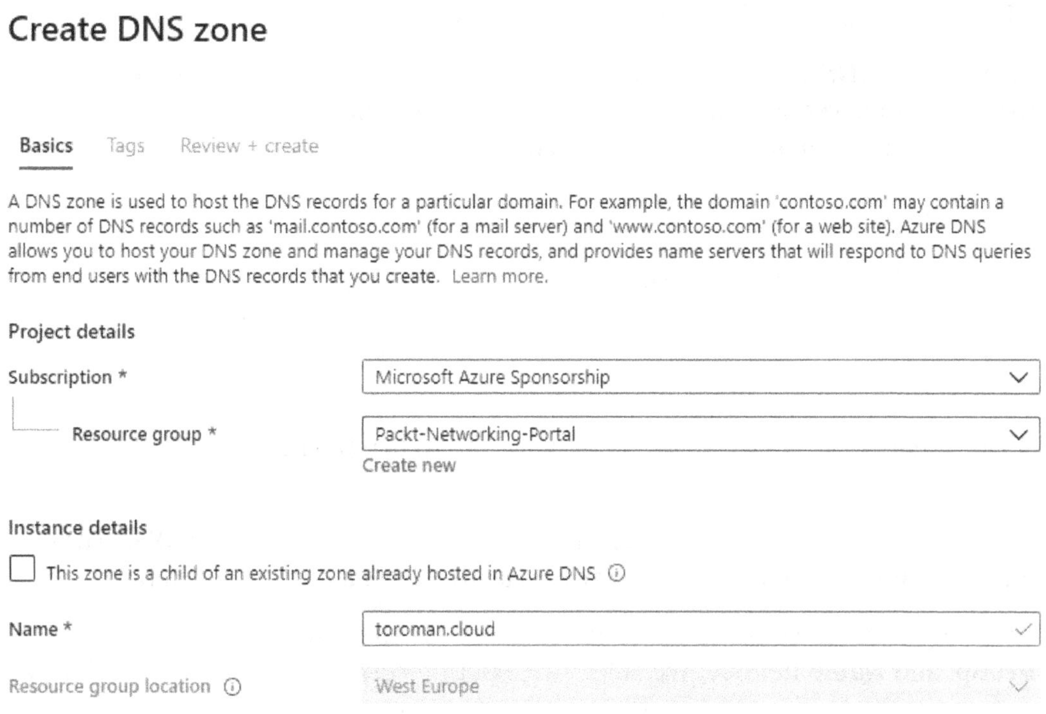

## Create DNS zone

Basics    Tags    Review + create

A DNS zone is used to host the DNS records for a particular domain. For example, the domain 'contoso.com' may contain a number of DNS records such as 'mail.contoso.com' (for a mail server) and 'www.contoso.com' (for a web site). Azure DNS allows you to host your DNS zone and manage your DNS records, and provides name servers that will respond to DNS queries from end users with the DNS records that you create.  Learn more.

**Project details**

| | |
|---|---|
| Subscription * | Microsoft Azure Sponsorship ⌄ |
| └── Resource group * | Packt-Networking-Portal ⌄ |
| | Create new |

**Instance details**

☐ This zone is a child of an existing zone already hosted in Azure DNS ⓘ

| | |
|---|---|
| Name * | toroman.cloud ✓ |
| Resource group location ⓘ | West Europe ⌄ |

**Figure 6.1: Creating a new Azure DNS zone with the Azure portal**

# How it works...

A DNS zone is required to start using Azure DNS. A new DNS zone is required for each domain we want to host with Azure DNS, as a single DNS zone can hold information for a single domain. After we create a DNS zone, we can add records, record sets, and route tables to a domain hosted with Azure DNS. Using these, we can route traffic and define destinations using an FQDN for Azure resources (and other resources as well). We'll show how to create and manage these in the coming recipes in this chapter.

Let's move on to the next recipe to learn how to create a private DNS zone.

# Creating an Azure Private DNS zone

An Azure Private DNS zone operates very similarly to a DNS zone. However, instead of operating on public records, it operates inside a virtual network. It is used to resolve custom names and domains inside your Azure virtual network.

## Getting ready

Before you start, open your browser and go to the Azure portal at https://portal.azure.com.

## How to do it...

In order to create a new Azure DNS zone with the Azure portal, we must follow these steps:

1. In the Azure portal, select **Create a resource** and choose **Private DNS Zone** under **Networking** services (or search for `Private DNS Zone` in the search bar).

2. In the new pane, we must enter information for the **Subscription**, **Resource group**, and **Name** fields. If we select an existing resource group, the region will automatically be the same as the one for the resource group selected. The name must be an FQDN:

### Create DNS zone

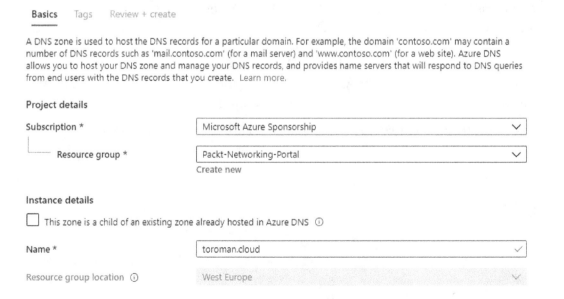

Figure 6.2: Creating a new private DNS zone with the Azure portal

## How it works...

When a virtual network is created, a default DNS zone is provided. The default DNS zone uses Azure-provided names, and we must use a private DNS zone to use custom names. A private DNS zone is also required for name resolution across virtual networks, as default DNS doesn't support such an option.

Let's move on to the next recipe to learn how to integrate a virtual network with a private DNS zone.

# Integrating a virtual network with a private DNS zone

When a private DNS zone is created, it is a standalone service that doesn't do much on its own. We must integrate it with a virtual network in order to start using it. Once integrated, it will provide DNS inside the virtual network.

## Getting ready

Before you start, open the browser and go to the Azure portal at https://portal.azure.com.

## How to do it...

In order to add a new record to the DNS zone, we must use the following steps:

1. In the Azure portal, locate **Private DNS Zone**.

2. In **Private DNS Zone**, select **Virtual network links** and click **Add**:

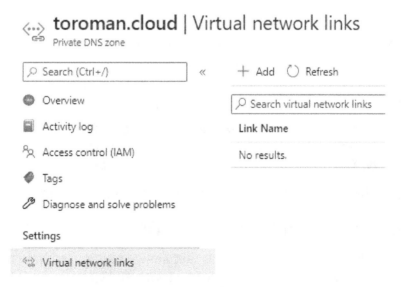

Figure 6.3: Adding a virtual network link

3. In the new pane, fill in **Link name**, then select values for the **Subscription** and **Virtual network** fields (only virtual networks in the selected subscription will be available). Alternatively, we can provide the resource ID of our virtual network, rather than selecting options from the drop-down menu:

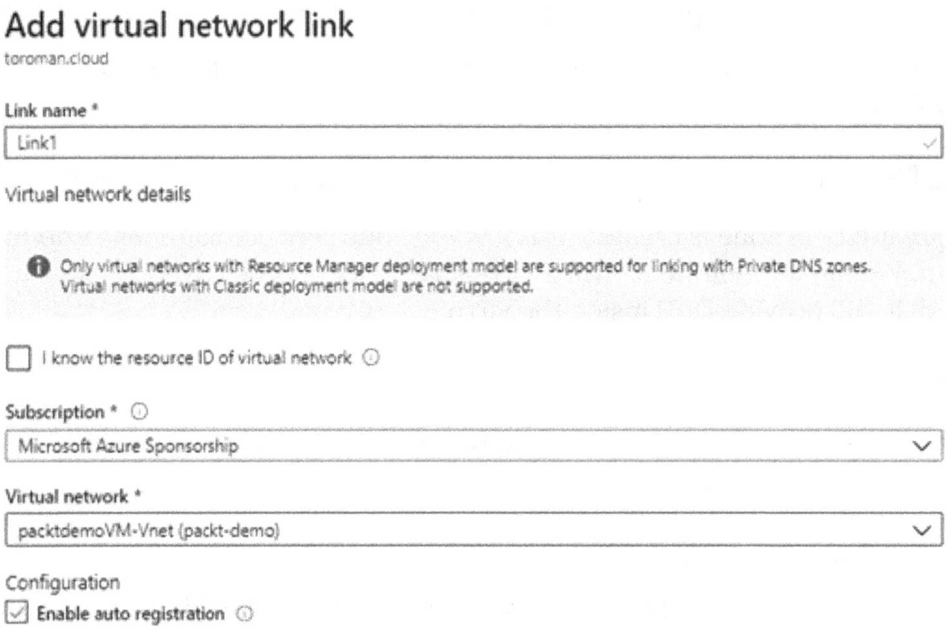

## Add virtual network link

toroman.cloud

Link name *

| Link1 | ✓ |

Virtual network details

> ℹ Only virtual networks with Resource Manager deployment model are supported for linking with Private DNS zones. Virtual networks with Classic deployment model are not supported.

☐ I know the resource ID of virtual network ⓘ

Subscription * ⓘ

| Microsoft Azure Sponsorship | ∨ |

Virtual network *

| packtdemoVM-Vnet (packt-demo) | ∨ |

Configuration
☑ Enable auto registration ⓘ

Figure 6.4: Adding a virtual network link

## How it works...

Once the virtual network is linked to the private DNS zone, the zone can be used for name resolution inside the connected virtual network. For name resolution across multiple connected virtual networks, we must use a private DNS zone, as default DNS doesn't support resolution across networks. The same applies if the network is connected to an on-premises network.

If we enable auto-registration under **Configuration**, newly created virtual machines will be automatically registered in the private DNS zone. Otherwise, we must add each new resource manually.

Let's move on to the next recipe to learn how to create a new record set in Azure DNS.

## Creating a new record set in Azure DNS

When creating a DNS zone, we define what domain we're going to hold records for. A DNS zone is created for a **root** domain defined with an FQDN. We can add additional subdomains and add records to hold information on other resources on the same domain.

## Getting ready

Before you start, open the browser and go to the Azure portal via https://portal.azure.com.

## How to do it...

In order to add a new record to the DNS zone, we must use the following steps:

1. In the Azure portal, locate **DNS zone**.

2. In **Overview**, select the option for adding a record set:

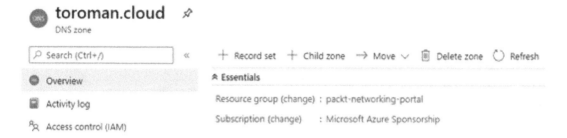

Figure 6.5: Adding a record set in DNS zone

3. A new pane will open. Enter the name of the subdomain to which you want to add a record:

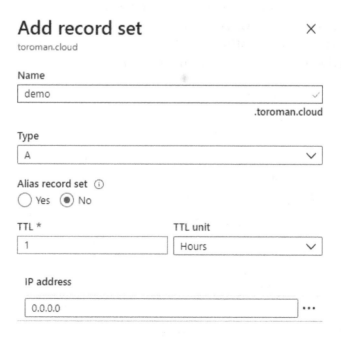

Figure 6.6: Adding a subdomain for the record

4. We need to select the type of record we want to add. The options are **A**, **AAAA**, **CNAME**, **MX**, **NS**, **SRV**, **TXT**, and **PTR**. The most common record type is **A**, so let's select that one:

**Figure 6.7: Selecting the record type**

5. After we select the record type, we need to select an alias (aliases are available only for the **A**, **AAAA**, and **CNAME** types) and the **TTL (Time-To-Live)** option. Finally, we add a record destination. This depends on the record type, and in the case of record **A**, it's going to be an IP address:

**Figure 6.8: Adding an alias, TTL, and record destination**

6.  If we choose **CNAME** as the record type, we are not entering an IP address but an alias. When a query is made for the record, instead of an IP address, a URL is returned and the client is directed to this record:

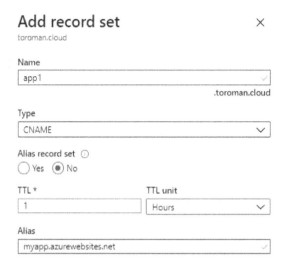

Figure 6.9: Adding a CNAME record

7.  Adding a single entry to our record creates a new record set and a new record. We can add more records to the record set by adding additional IP addresses (in this case).

## How it works...

A DNS record set holds information on the subdomain in the domain hosted with the DNS zone. In this case, the domain would be **toroman.cloud**, and the subdomain would be **test**. This forms an FQDN, **demo.toroman.cloud**, and the record points this domain to the IP address we defined. The record set can hold multiple records for a single subdomain, usually used for redundancy and availability.

Using **CNAME** and/or an alias can be done with Azure Traffic Manager. In this way, custom domain names can be used for name resolution, instead of the default names provided by Azure.

In this recipe, you learned how to create a new record in Azure DNS. Let's move on to the next recipe to learn how to create a route table.

## Creating a route table

Azure routes network traffic in subnets by default. However, in some cases, we want to use custom traffic routes to define where and how traffic flows. In such cases, we use **route tables**. A route table defines the next hop for our traffic and determines where the network traffic needs to go.

## Getting ready

Before you start, open the browser and go to the Azure portal via https://portal.azure.com.

## How to do it...

In order to add a new record to the DNS zone, we must use the following steps:

1. In the Azure portal, select **Create a resource** and choose **Route Table** under **Networking** services (or search for **route table** in the search bar).

2. In the new pane, we need to select options for **Subscription**, **Resource group**, and **Region**, and provide the name of the route table. Optionally, we can define whether we want to allow gateway route propagation (which is enabled by default):

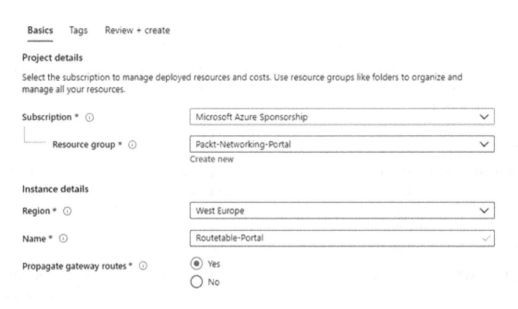

Figure 6.10: Creating a route table

## How it works...

Network routing in Azure Virtual Network is done automatically, but we can use custom routing with route tables. Route tables use rules and subnet associations to define traffic flow in Virtual Network. When a new route table is created, no configuration is created—only an empty resource. After the resource is created, we need to define rules and subnets in order to use a route table for the traffic flow. We will show in the coming recipes in this chapter how we create and apply rules in route tables.

# Changing a route table

As mentioned in the previous recipe, creating a new route table will result in an empty resource. Once a resource is created, we can change the settings as needed. Before we configure the routes and subnets associated with the route table, the only setting we can change is the **Border Gateway Protocol** (**BGP**) route propagation. We may change other settings after creation as well.

## Getting ready

Before you start, open the browser and go to the Azure portal via https://portal.azure. com.

## How to do it...

In order to change a route table, we must do the following:

1.  In the Azure portal, locate **Route table**.

2.  Under **Settings**, we may change the **Propagate gateway routes** settings in the **Configuration** pane at any time:

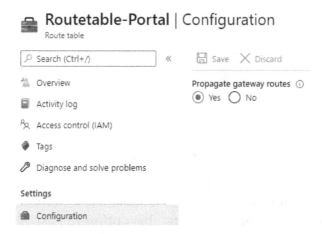

Figure 6.11: Option to change the Propagate gateway routes settings

## How it works...

Under the settings of the route table, we can disable or enable gateway route propagation at any time. This option, if disabled, prevents on-premises routes from being propagated via BGP to the network interfaces in a virtual network subnet. Under the settings, we can create, delete, or change routes and subnets. These options will be addressed in the coming recipes in this chapter.

Let's move on to the next recipe, where you will learn how to associate a route table with a subnet.

# Associating a route table with a subnet

When a route table is created, it doesn't do anything until it's properly configured. There are two things we need to address: which resources are affected, and how. To define which resources are affected, we must make an association between a subnet and a route table.

## Getting ready

Before you start, open the browser and go to the Azure portal via https://portal.azure.com.

## How to do it...

In order to associate a subnet with a route table, we must do the following:

1.  In the Azure portal, locate **Route table**.

2.  Under **Settings**, select the **Subnets** option. In the **Subnets** pane, select the **Associate** option to create a new association:

Figure 6.12: Creating a new association

3.  A new pane will open. There are two options available—selecting a virtual network and choosing a subnet that we want to associate the route table with. First, we must select **Virtual network**. Selecting this option will list all the available virtual networks. Select the one you want to associate from this list:

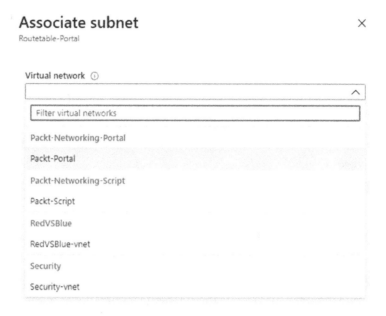

**Figure 6.13: Selecting a virtual network**

4.  After a virtual network is selected, we can proceed to select a subnet. The **Subnet** option will list all the subnets from the virtual network we selected in the previous step. Choose the subnet you want to associate from this list:

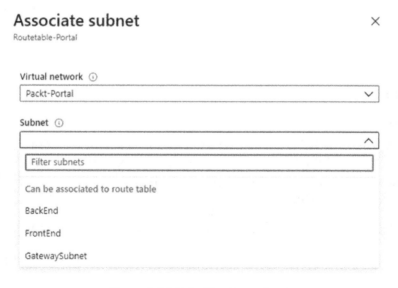

**Figure 6.14: Selecting the subnet**

5.  After both options are selected, we may proceed to create an association:

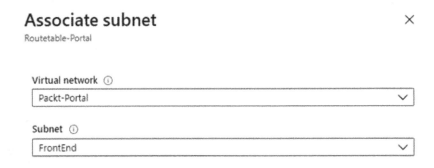

**Figure 6.15: Virtual network and subnet selected**

6.  After a subnet has been associated, it will appear in a list of subnets under the route table:

**Figure 6.16: List of associated subnets**

## How it works...

The route table, to be effective, must have two parts defined: the *what* and the *how*. We define what will be affected by the route table with a subnet association. This is only one part of the configuration, as just associating a subnet to a route table will do nothing. We must create rules that will apply to this association. We'll explain the rules in the following recipes in this chapter.

Let's move on to a new recipe and learn how to dissociate a route table from a subnet.

# Dissociating a route table from a subnet

After we create an association and rules, those rules will apply to all resources on the associated subnet. If we want rules to no longer apply to a specific subnet, we can remove the association.

## Getting ready

Before you start, open the browser and go to the Azure portal via https://portal.azure.com.

## How to do it...

In order to remove the association between the subnet and the route table, we must do the following:

1.  In the Azure portal, locate **Route table**.

2.  Under **Settings**, select the **Subnets** option, and select the subnet you want to remove:

Figure 6.17: Selecting a subnet for removal

3. The subnet configuration pane will open. Select the **Route table** option. Note that this actually opens a subnet configuration. It's a common mistake to confuse this pane with the association and to choose the **Delete** option. This will not only remove the association but also remove the subnet altogether:

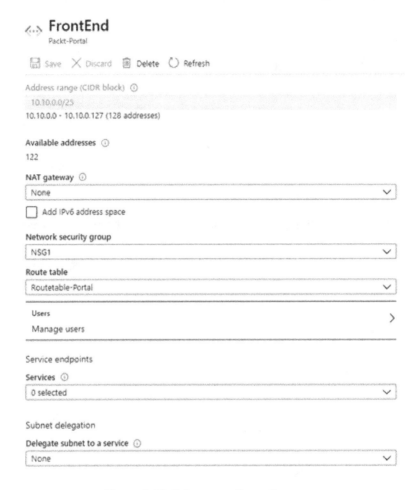

Figure 6.18: Subnet configuration pane

4. The Azure portal will show a list of the available route tables for a specific subnet. Choose **None**:

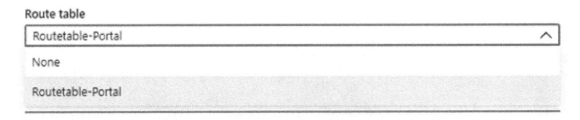

Figure 6.19: List of available route tables for a subnet

5. After selecting **None**, click the **Save** button to apply the new settings. The route table association is removed from the subnet:

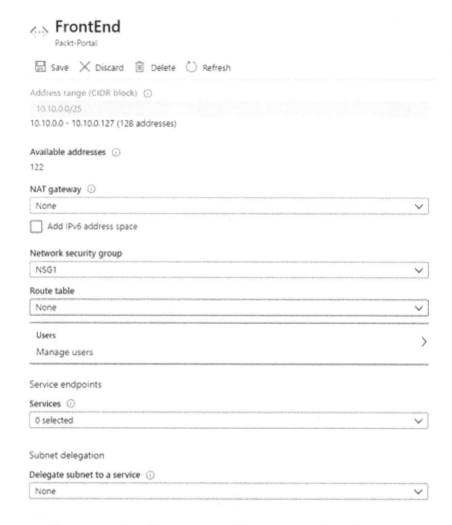

Figure 6.20: Removing a route table association from the subnet

## How it works...

At some point, we may have created rules in a route table that apply to multiple subnets. If we no longer want to apply one or more rules to a specific subnet, we can remove the association. Once the association is removed, the rules will no longer apply to the subnet. All rules will apply to all the associated subnets. If we need to make a single rule no longer apply to a specific subnet, we must remove the association.

In this recipe, we learned how to dissociate a route table. Let's move on to the next recipe and learn how to create a new route.

# Creating a new route

After we create a route table and the associated subnets, there is still a piece missing. We defined the route table that will be affected with subnet association, but we're missing the part that defines *how* it will be affected. We define how associated subnets are affected using rules called **routes**. Routes define traffic routes, stating where specific traffic needs to go. If the default route for specific traffic is the internet, we can change this and reroute the traffic to a specific IP or subnet.

## Getting ready

Before you start, open the browser and go to the Azure portal via https://portal.azure.com.

## How to do it...

In order to create a new route, we must do the following:

1.  In the Azure portal, locate **Route table**.

2.  In the **Route table** pane, under **Settings**, select **Routes**. Select **Add** to add a new route:

Figure 6.21: Adding a new route

3.  In the new pane, we need to define values for the **Route name** and **Address prefix** (in CIDR format) fields for the destination IP address range and select an option for **Next hop type**. The options for this include **Virtual network gateway**, **Virtual network**, **Internet**, **Virtual appliance**, and **None**:

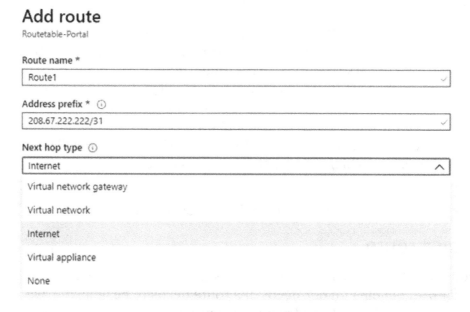

Figure 6.22: Adding route details

4. The last option, **Next hop address**, is active only when a virtual appliance is used. In that case, we need to provide the virtual appliance IP address in this field, and all traffic will go through the virtual appliance. Let's choose **Internet** and provide a public IP address in the **Address prefix** field (the **Address prefix** option always depends on the **Next hop type** option):

Figure 6.23: Selecting Internet for Next hop type

## How it works...

The route defines the traffic flow. All traffic from the associated subnet will follow the route defined by these rules. If we define that traffic will go to the internet, all traffic will go outside the network to an IP address range defined with an IP address prefix. If we choose for traffic to go to a virtual network, it will go to a subnet defined by the IP address prefix. If that virtual network gateway is used, all traffic will go through the virtual network gateway and reach its connection on the other side—either another virtual network or our local network. The **Virtual appliance** option will send all traffic to the virtual appliance, which then, with its own set of rules, defines where the traffic goes next.

Let's move on to the next recipe and learn how to change a route.

# Changing a route

Route requirements may change over time. In such cases, we can either remove the route or edit it, depending on our needs. If a route needs to be adjusted, we can select the option to change the route and apply the new traffic flow at any time.

## Getting ready

Before you start, open the browser and go to the Azure portal via https://portal.azure. com.

## How to do it...

In order to change the existing route, we need to do the following:

1.  In the Azure portal, locate **Route table**.

2.  Under **Settings**, select **Routes** and select the route you want to change from the list of available routes:

Figure 6.24: Changing an available route

3. A new pane will open. We can change the **Address prefix** (for the destination IP range) and **Next hop type** settings. If the **Next hop type** option is a virtual appliance, an option for **Next hop address** will be available:

**Figure 6.25: Option for Next hop address**

## How it works...

The requirements for a route may change over time. We can change a route and adjust it to suit new requirements as needed. The most common scenarios are that the traffic needs to reach a specific service when the service IP changes over time. For example, we may need to route all traffic through a virtual appliance, but the IP address of the virtual appliance changes over time. We may change the route in the route table to reflect this change and force the traffic flow through the virtual appliance. Another example is when traffic needs to reach our local network through a virtual network gateway–the destination IP address range may change over time, and we need to reflect these changes in the route once again.

In this recipe, we learned how to change a route. In the next recipe, we'll learn how to delete a route.

# Deleting a route

As we have already mentioned, route requirements may change over time. In some cases, rules are no longer applicable and we must remove them. In such cases, changing the route will not complete the task and we will need to remove the route completely. This task may be completed by deleting the route.

## Getting ready

Before you start, open the browser and go to the Azure portal via https://portal.azure.com.

## How to do it...

In order to delete a route, we must do the following:

1.  In the Azure portal, locate the **Route table** link.

2.  Under **Settings**, select **Routes** and then select the route you want to delete:

Figure 6.26: Deleting an existing route

3.  A new pane will open. Select the **Delete** option and confirm your action:

Figure 6.27: Selecting the Delete option

4.  After this action has been confirmed, you will return to the previous pane and the deleted route will no longer be listed:

Figure 6.28: The successful deletion of a route

## How it works...

As our requirements change, we need to address the new requirements in our route tables. We can either edit routes or remove them to meet these new requirements. When multiple routes are used in a single route table, one of the routes may become obsolete, or even block new requirements. In such cases, we may want to delete a route to resolve any issues.

# Azure Firewall

Most Azure networking components used for security are there to stop unwanted incoming traffic. Whether we use network security groups, application security groups, or a **Web Application Firewall (WAF)**, they all have one single purpose—to stop unwanted traffic from reaching our services. Azure Firewall has similar functionality, including one extension that we can use to stop outbound traffic from leaving the virtual network.

We will cover the following recipes in this chapter:

- Creating a new firewall
- Creating a new firewall with PowerShell
- Configuring a new allow rule
- Configuring a new deny rule
- Configuring a route table
- Enabling diagnostic logs for Azure Firewall
- Configuring Azure Firewall in forced tunneling mode
- Creating an IP group
- Configuring Azure Firewall DNS settings

# Technical requirements

For this chapter, the following is required:

- An Azure subscription
- Azure PowerShell

The code samples can be found at https://github.com/PacktPublishing/Azure-Networking-Cookbook-Second-Edition/tree/master/Chapter07.

# Creating a new firewall

Azure Firewall gives us total control over our traffic. Besides controlling inbound traffic, with Azure Firewall, we can control outbound traffic as well.

## Getting ready

Before we can create an Azure Firewall instance, we must first prepare a subnet.

In order to create a new subnet for Azure Firewall, we must do the following:

1. Locate the virtual network that will be associated with our Azure Firewall.

2. Select the **Subnets** option under **Settings** and click **Subnet** to add a new subnet, as shown in *Figure 7.1*:

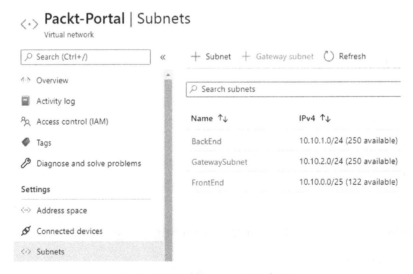

**Figure 7.1: Adding a new subnet**

3. In the new pane, we must provide values for the **Name** and **Address range** fields. It's very important that the subnet is named `AzureFirewallSubnet`:

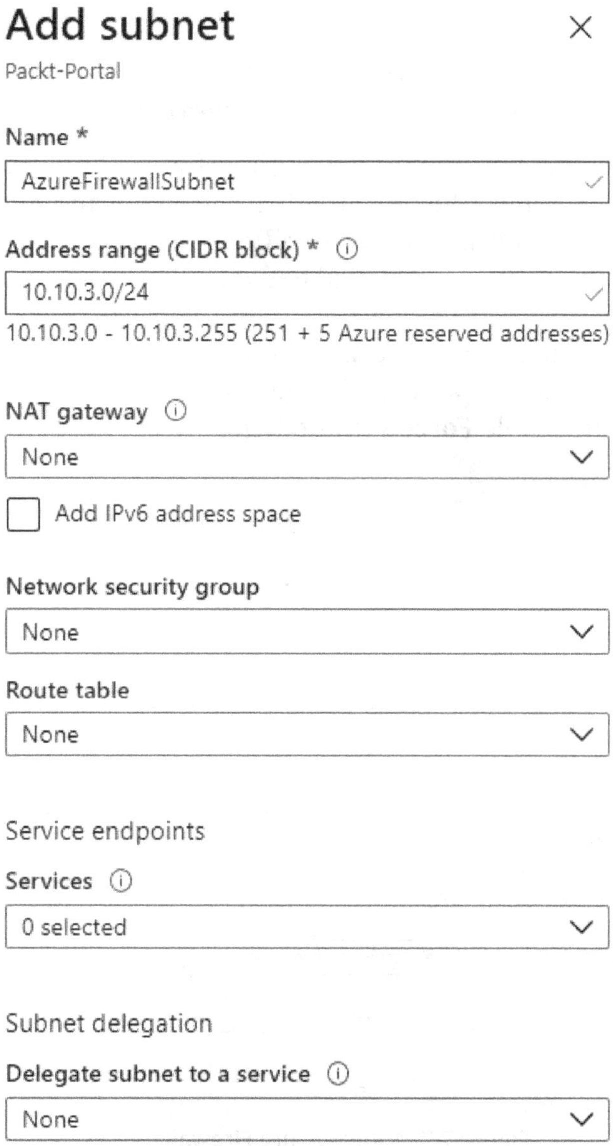

## Add subnet  ✕
Packt-Portal

Name *

| AzureFirewallSubnet | ✓ |

Address range (CIDR block) *  ⓘ

| 10.10.3.0/24 | ✓ |

10.10.3.0 - 10.10.3.255 (251 + 5 Azure reserved addresses)

NAT gateway ⓘ

| None | ⌄ |

☐ Add IPv6 address space

Network security group

| None | ⌄ |

Route table

| None | ⌄ |

Service endpoints

Services ⓘ

| 0 selected | ⌄ |

Subnet delegation

Delegate subnet to a service ⓘ

| None | ⌄ |

Figure 7.2: Providing the name and address range of the subnet

## How to do it...

In order to create a new Azure Firewall instance using the Azure portal, take the following steps:

1. In the Azure portal, select **Create a resource** and choose **Azure Firewall** under **Networking** services (or search for `Azure Firewall` in the search bar).

2. In the new pane, first, we must provide values for the **Subscription** and **Resource group** drop-down menus. We need to fill in the **Name** and **Region** fields for Azure Firewall, and optionally select an **Availability zone** option. Next, we proceed to virtual network selection. Only virtual networks in the region where the Azure Firewall instance will be created are available. Also, the selected virtual network must contain the `AzureFirewallSubnet` subnet we created earlier. Finally, we define a public IP address (we can choose an existing one or create a new one). Optionally, we can enable **Forced tunneling**:

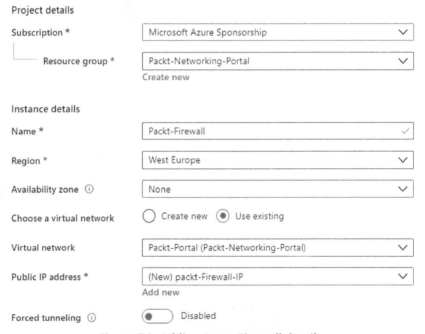

**Figure 7.3: Adding Azure Firewall details**

## How it works...

Azure Firewall uses a set of rules to control outbound traffic. We can either block everything by default and allow only whitelisted traffic, or we can allow everything and block only blacklisted traffic. It's essentially the central point where we can set network policies, enforce these policies, and monitor network traffic across virtual networks or even subscriptions. As a firewall as a service, Azure Firewall is a managed service with built-in high availability and scalability.

# Creating a new firewall with PowerShell

Alternatively, we can deploy Azure Firewall using PowerShell. This method is especially useful when services are part of a large deployment or any deployment that needs to be automated.

## How to do it...

There are several steps that need to be executed in order to create a new firewall with Azure PowerShell:

1. First, we define the parameters:

```
$RG="Packt-Networking-Script"
$Location="West Europe"
$VNetName = "Packt-Script"
$AzFwIpName = "AzFW-Public-IP"
$AzFwname = "AzFw-Script"
```

2. Then, we need to create a separate subnet for Azure Firewall:

```
$vnet = Get-AzVirtualNetwork -ResourceGroupName $RG '
-Name $VnetName
Add-AzVirtualNetworkSubnetConfig -Name AzureFirewallSubnet '
-VirtualNetwork $vnet '
-AddressPrefix 10.11.3.0/24
Set-AzVirtualNetwork -VirtualNetwork $vnet
```

3. Next, we need to create a public IP address for Azure Firewall:

```
$AzFwIp = New-AzPublicIpAddress -Name $AzFwIpName  '
-ResourceGroupName $RG '
-Location $Location '
-AllocationMethod Static '
-Sku Standard
```

4. Finally, we have all the components in place and can proceed to create the firewall:

```
$Azfw = New-AzFirewall -Name $AzFwname '
-ResourceGroupName $RG '
-Location $Location '
-VirtualNetworkName $vnet.Name '
-PublicIpName $AzFwIp.Name
```

## How it works...

The firewall requires a separate subnet that is named **AzureFirewallSubnet**. So, we need to create such a subnet on the virtual network we intend to use. Another requirement is a public IP address. Finally, we are ready for deployment and can create a new Azure Firewall instance.

But deploying Azure Firewall is just the start. We need to configure our firewall by creating rules and routes. Let's proceed to the next recipe and see how rules are created.

# Configuring a new allow rule

If we want to allow specific traffic, we must create an allow rule. Rules are applied based on priority level, so a rule will be applied only when there is no other rule with higher priority.

## Getting ready

Open the PowerShell console and make sure you are connected to your Azure subscription.

## How to do it...

In order to create a new allow rule in Azure Firewall, execute the following command:

```
$RG="Packt-Networking-Script"

$Location="West Europe"

$Azfw = Get-AzFirewall -ResourceGroupName $RG

$Rule = New-AzFirewallApplicationRule -Name Rule1 -Protocol
"http:80","https:443" -TargetFqdn "*packt.com"

$RuleCollection = New-AzFirewallApplicationRuleCollection -Name
RuleCollection1 -Priority 100 -Rule $Rule -ActionType "Allow"

$Azfw.ApplicationRuleCollections = $RuleCollection

Set-AzFirewall -AzureFirewall $Azfw
```

## How it works...

An allow rule in Azure Firewall will whitelist specific traffic. If there is a rule that would also block this traffic, the higher-priority rule will be applied.

We can create deny rules as well. Let's see how we can do that in the next recipe.

# Configuring a new deny rule

If we want to deny specific traffic, we must create a deny rule. Rules are applied by priority, so this rule will be applied only if there is not a higher-priority rule in effect.

## Getting ready

Open the PowerShell console and make sure you are connected to your Azure subscription.

## How to do it...

In order to create a new deny rule in Azure Firewall, execute the following command:

```
$RG="Packt-Networking-Script"

$Location="West Europe"

$Azfw = Get-AzFirewall -ResourceGroupName $RG

$Rule = New-AzFirewallApplicationRule -Name Rule1 -Protocol
"http:80","https:443" -TargetFqdn "*google.com"

$RuleCollection = New-AzFirewallApplicationRuleCollection -Name
RuleCollection1 -Priority 100 -Rule $Rule -ActionType "Deny"

$Azfw.ApplicationRuleCollections = $RuleCollection

Set-AzFirewall -AzureFirewall $Azfw
```

## How it works...

The deny rule is the most commonly used option with Azure Firewall. An approach where you block everything and allow only whitelisted traffic isn't very practical, as we may end up adding a great many allow rules. Therefore, the most common approach is to use deny rules to block certain traffic that we want to prevent.

# Configuring a route table

Route tables are commonly used with Azure Firewall when there is cross-connectivity. Cross-connectivity can either be with other Azure virtual networks or with on-premises networks. In such cases, Azure Firewall uses route tables to forward traffic based on the rules specified in the route tables.

## Getting ready

Open the PowerShell console and make sure you are connected to your Azure subscription.

## How to do it...

In order to create a new route table in Azure Firewall, execute the following command:

```
$RG="Packt-Networking-Script"

$Location="West Europe"

$Azfw = Get-AzFirewall -ResourceGroupName $RG

$config = $Azfw.IpConfigurations[0].PrivateIPAddress

$Route = New-AzRouteConfig -Name 'Route1' -AddressPrefix 0.0.0.0/0 -NextHopType
VirtualAppliance -NextHopIpAddress $config

$RouteTable = New-AzRouteTable -Name 'RouteTable1' -ResourceGroupName $RG
-location $Location -Route $Route
```

## How it works...

Using route tables associated with Azure Firewall, we can define how traffic between networks is handled and how we route traffic from one network to another. In a multi-network environment, especially in a hybrid network where we connect an Azure virtual network with a local on-premises network, this option is very important. This allows us to determine what kind of traffic can flow where and how.

# Enabling diagnostic logs for Azure Firewall

Diagnostics are a very important part of any IT system, and networking is no exception. The diagnostics settings in Azure Firewall allow us to collect various information that can be used for troubleshooting or auditing.

## Getting ready

Before you start, open your browser and go to the Azure portal at https://portal.azure.com.

## How to do it...

To enable diagnostics in Azure Firewall, we must follow these steps:

1. In the Azure Firewall pane, locate **Diagnostics settings** under **Monitoring**.

2. Select the **Add diagnostic setting** option, as shown in *Figure 7.4*:

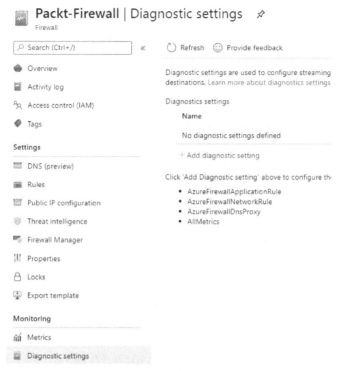

**Figure 7.4: Adding a diagnostic setting**

3.  In the new pane, fill in the name field and specify where the logs will be stored. Choose the storage account where the logs will be stored, and specify the retention period and which logs will be stored, as shown in *Figure 7.5*:

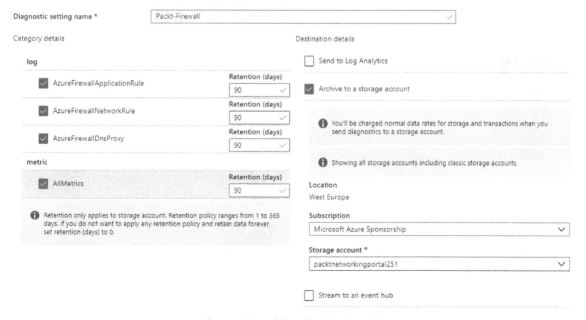

**Figure 7.5: Adding the log details**

## How it works...

Diagnostics has two purposes—auditing and troubleshooting. Based on traffic and settings, these logs can grow over time, so it's important to consider the main purpose of enabling diagnostics in the first place. If diagnostics are enabled for auditing, you will probably want to choose a maximum of 365 days of retention. If the main purpose is troubleshooting, the retention period can be kept at 7 days or an even shorter period of time. Setting the retention policy to **0** will store logs without removing them after a period of time. This can generate additional costs and you may need to set up a different procedure for removing logs.

If we don't want to store diagnostic logs in a storage account, we can choose Log Analytics or Event Hubs. The process, in this case, does not include setting retention periods as these settings are kept on the destination side.

# Configuring Azure Firewall in forced tunneling mode

Forced tunneling allows us to force all internet-bound traffic to an on-premises firewall for inspection or audit. Because of different Azure dependencies, this is not enabled by default and requires User Defined Routes (USRs) to allow forced tunneling. This is also not possible by using `AzureFirewallSubnet`, and we need to add an additional subnet named `AzureFirewallManagementSubnet`. Note that this needs to be done prior to Azure Firewall deployment and will not work if the subnet is added afterward.

## Getting ready

Before you start, open your browser and go to the Azure portal at https://portal.azure. com.

## How to do it...

In order to add `AzureFirewallManagementSubnet` for forced tunneling, we need to do the following:

1.  In the Azure portal, select **Create a resource** and choose **Route Table** under **Networking** services (or search for `Route Table` in the search bar).

2. In the new pane, provide information for the **Subscription**, **Resource group**, **Region**, and **Name** fields for the route table. Make sure to select **No** for **Propagate gateway routes**:

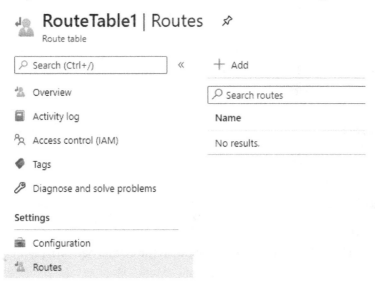

## Create Route table

Basics    Tags    Review + create

**Project details**

Select the subscription to manage deployed resources and costs. Use resource groups like folders to organize and manage all your resources.

Subscription *   ⓘ      Microsoft Azure Sponsorship

     Resource group *   ⓘ      Packt-Portal

                            Create new

**Instance details**

Region *   ⓘ      West Europe

Name *   ⓘ      RouteTable1

Propagate gateway routes *   ⓘ     ○ Yes
                                                   ◉ No

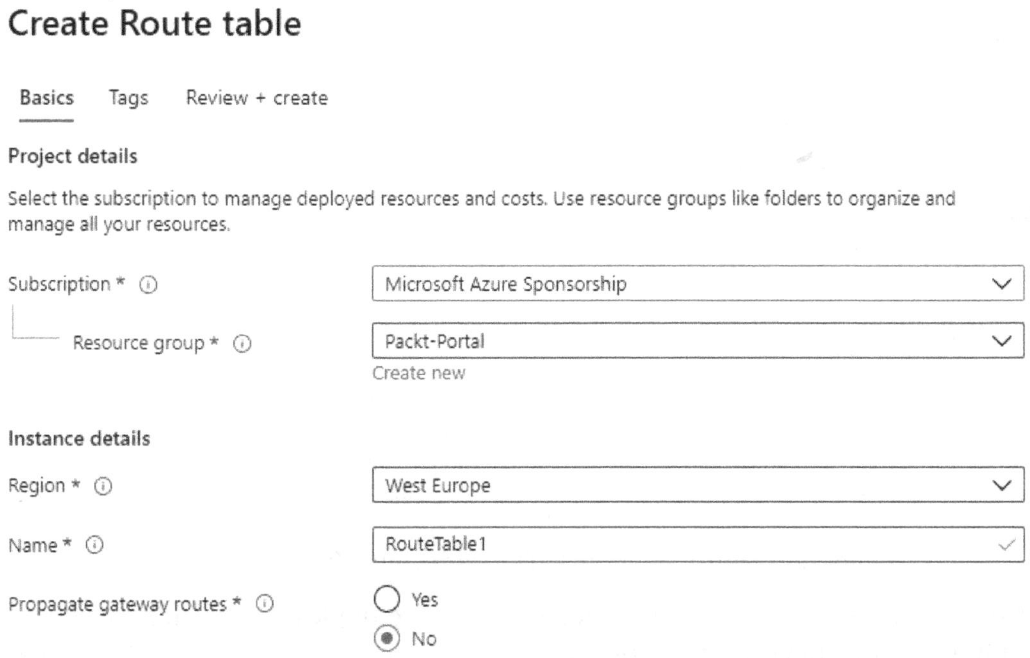

**Figure 7.6: Creating a route table using the Azure portal**

3. Once the route table is created, we need to set a default internet route. Go to the route table we just created, and under **Routes** in the **Settings** section, select **Add**:

### RouteTable1 | Routes
Route table

🔍 Search (Ctrl+/)     «     ➕ Add

   Overview                  🔍 Search routes

   Activity log                Name

   Access control (IAM)       No results.

   Tags

   Diagnose and solve problems

**Settings**

   Configuration

   Routes

**Figure 7.7: Adding a default internet route for the route table**

4. In the new pane, we need to provide a name for the route. We should also put `0.0.0.0/0` under **Address prefix** and `Internet` under **Next hop type**:

# Add route
RouteTable1

Route name *

    Internet                                                                    ✓

Address prefix * ⓘ

    0.0.0.0/0                                                                   ✓

Next hop type ⓘ

    Internet                                                                    ⌄

Next hop address ⓘ

Figure 7.8: Configuring the default internet route for the route table

5. Now go to the virtual network where you plan to deploy Azure Firewall. Under **Subnets**, add a new subnet. Note that `AzureFirewallSubnet` still needs to be added as well:

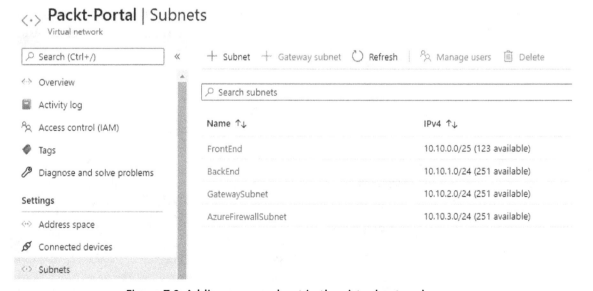

Figure 7.9: Adding a new subnet in the virtual network pane

6. In the new pane, set the name to **AzureFirewallManagementSubnet**, provide a value for the **Subnet address range** field (a minimum subnet size of **/26** is required), and select the route table we created in the **Route table** field:

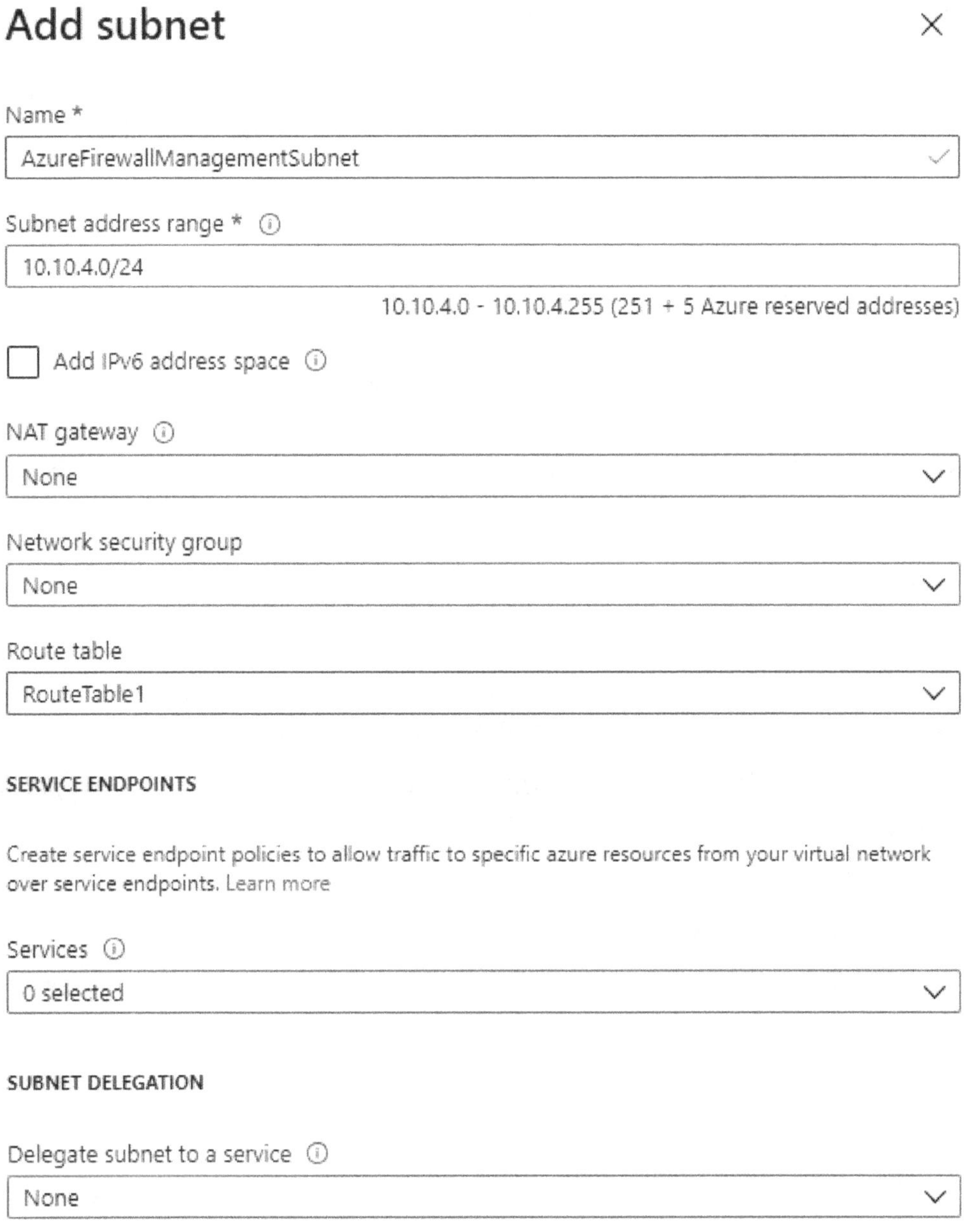

# Add subnet                                                    ✕

Name *

| AzureFirewallManagementSubnet                              ✓ |

Subnet address range * ⓘ

| 10.10.4.0/24                                                 |

10.10.4.0 - 10.10.4.255 (251 + 5 Azure reserved addresses)

☐ Add IPv6 address space ⓘ

NAT gateway ⓘ

| None                                                       ∨ |

Network security group

| None                                                       ∨ |

Route table

| RouteTable1                                                ∨ |

**SERVICE ENDPOINTS**

Create service endpoint policies to allow traffic to specific azure resources from your virtual network over service endpoints. Learn more

Services ⓘ

| 0 selected                                                 ∨ |

**SUBNET DELEGATION**

Delegate subnet to a service ⓘ

| None                                                       ∨ |

Figure 7.10: Configuring the subnet settings in the new pane

7. Now we can proceed with Azure Firewall deployment. See the *Creating a new firewall* recipe.

## How it works...

In order to support forced tunneling, traffic associated with service management is separated from the rest of the traffic. An additional subnet is required with a minimum size of **/26**, along with an associated public IP address. A route table is required with a single route defining the route to the internet, and **BGP route propagation (propagate gateway routes)** must be disabled. We can now include routes and define where exactly traffic needs to go (a virtual network appliance or on-premises firewall) in order to be inspected or audited before reaching the internet.

# Creating an IP group

IP groups are Azure resources that help to group IP addresses for easier management. This way, we can apply Azure Firewall rules easier and with better visibility.

## Getting ready

Before you start, open your browser and go to the Azure portal at https://portal.azure.com.

## How to do it...

In order to create a new IP group, we need to do the following:

1. In the Azure portal, select **Create a resource** and choose **IP Group** under **Networking** services (or search for `IP group` in the search bar).

2. In the new pane, provide information for **Subscription**, **Resource group**, **Name**, and **Region**:

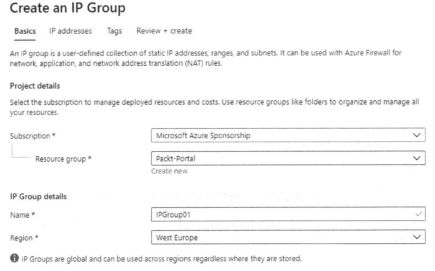

**Figure 7.11: Creating a new IP group using the Azure portal**

3. Under **IP addresses**, we need to provide something for the **IP address, range or subnet** field. In this example, we are adding a subnet:

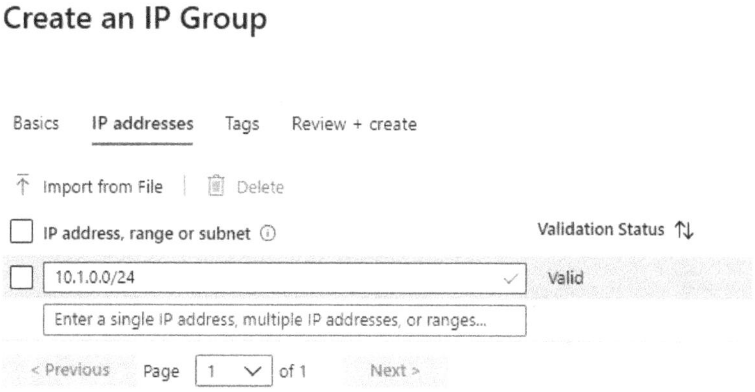

## Create an IP Group

Basics    **IP addresses**    Tags    Review + create

↑ Import from File    |    🗑 Delete

☐ IP address, range or subnet ⓘ                                    Validation Status ↑↓

☐ [ 10.1.0.0/24                                    ✓ ]    Valid

[ Enter a single IP address, multiple IP addresses, or ranges... ]

< Previous    Page [ 1 ∨ ] of 1    Next >

Figure 7.12: Adding a subnet in the IP address, range or subnet field

4. We can now proceed and deploy the IP group.

## How it works...

IP groups allow us to associate multiple IP addresses with a single resource for easier management. We can associate any number of individual IP addresses (in `10.10.10.10` format), IP ranges (in `10.10.10.10-10.10.10.20` format), or subnets (in `10.10.10.0/24` format). Then, firewall rules can be associated with IP groups and all IP addresses under a defined IP group. Instead of creating a separate rule for each IP address, range, or subnet, we can now have a single rule for a single IP range. This means easier management and maintenance of Azure Firewall, along with better visibility of effective rules.

# Configuring Azure Firewall DNS settings

We can use a custom DNS server with our Azure Firewall instance. This allows us to resolve custom names and apply filtering based on **Fully Qualified Domain Name (FQDN)**.

## Getting ready

Before you start, open your browser and go to the Azure portal at https://portal.azure.com.

## How to do it...

In order to configure custom DNS settings in Azure Firewall, we need to do the following:

1. In the Azure Firewall pane, locate **DNS** under **Settings**. We need to set it to **Enabled**. Select the type of DNS (default or custom) and whether we want to use a DNS proxy:

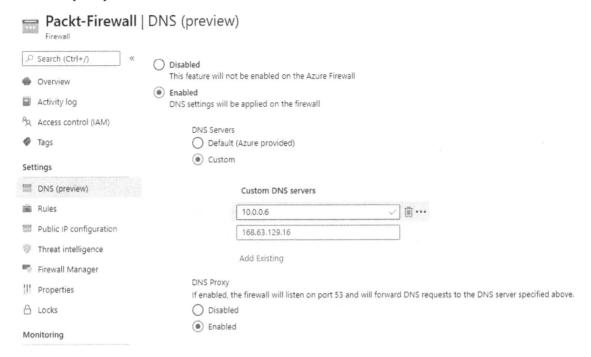

**Figure 7.13: Configuring Azure Firewall DNS settings using the Azure portal**

2. Once all the necessary settings are provided, select **Save** to apply them. It takes up to 30 minutes to correctly propagate routes and for them to take full effect.

## How it works...

In order to use FQDN filtering, Azure Firewall needs to be able to resolve the FQDN in question. This can be achieved by enabling DNS settings on Azure Firewall. When enabled, we can choose between Azure-provided DNS or custom DNS. Custom DNS can be either an Azure DNS zone or a DNS server running on a virtual network.

# 8

# Creating hybrid connections

Hybrid connections allow us to create secure connections with **Azure virtual networks (VNets)**. These connections can either be from on-premises or from other Azure VNets. Establishing connections to Azure VNets enables the exchange of secure network traffic with other services that are located in different Azure VNets, different subscriptions, or outside Azure (in different clouds or on-premises). Using secure connections removes the need for publicly exposed endpoints that present a potential security risk. This is especially important when we consider management, where opening public endpoints creates a security risk and presents a major issue. For example, if we consider managing virtual machines, it's a common practice to use either **Remote Desktop Protocol (RDP)** or PowerShell for management. Exposing these ports to public access presents a great risk. A best practice is to disable any kind of public access to such ports and use only access from an internal network for management. In this case, we use either a Site-to-Site or a Point-to-Site connection to enable secure management.

In another scenario, we might need the ability to access a service or a database on another network, either on-premises or via another Azure VNet. Again, exposing these services might present a risk, and we use either Site-to-Site, VNet-to-VNet, or VNet peering to enable such a connection in a secure way.

We will cover the following recipes in this chapter:

- Creating a Site-to-Site connection
- Downloading the VPN device configuration from Azure
- Creating a Point-to-Site connection
- Creating a VNet-to-VNet connection
- Connecting VNets using network peering

## Technical requirements

For this chapter, the following are required:

- An Azure subscription
- Windows PowerShell

The code samples can be found at https://github.com/PacktPublishing/Azure-Networking-Cookbook-Second-Edition/tree/master/Chapter08.

## Creating a Site-to-Site connection

A Site-to-Site connection is used to create a secure connection between an on-premises network and an Azure VNet. This connection is used to perform a number of different tasks, such as enabling hybrid connections or secure management. In a hybrid connection, we allow a service in one environment to connect to a service in another environment. For example, we could have an application in Azure that uses a database located in an on-premises environment. Secure management lets us limit management operations to being allowed only when coming from a secure and controlled environment, such as our local network.

## Getting ready

Before you start, open your browser and go to the Azure portal at https://portal.azure.com.

## How to do it...

To create a new Site-to-Site connection, we must follow these steps:

1. Locate the virtual network gateway (the one we created in *Chapter 5*, *Local and virtual network gateways*) and select **Connections**.

2. In **Connections**, select the **Add** option to add a new connection:

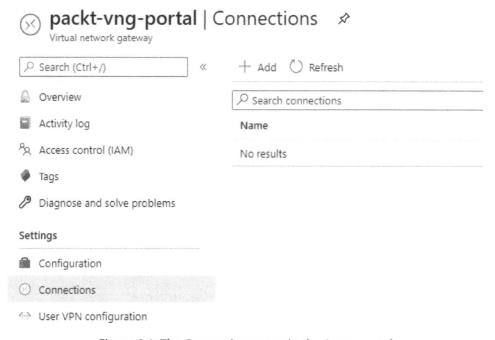

Figure 8.1: The Connections pane in the Azure portal

3. In the new pane, we need to enter the connection name and select **Site-to-site (IPsec)** for **Connection type**:

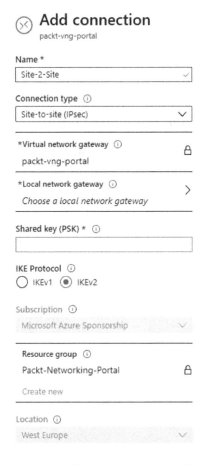

Figure 8.2: Adding connection attributes

4. Under **Local network gateway**, we need to select a local network gateway from the list (we created a local network gateway in *Chapter 5, Local and virtual network gateways*):

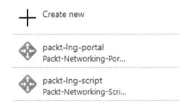

Figure 8.3: Selecting a local network gateway

5.  We need to provide a shared key in the **Shared key (PSK)** field that will be used for the IPSec connection. We also need to define the **IKE protocol** that will be used for security association. We can choose between **IKEv1** and **IKEv2**. Note that the options for **Subscription**, **Resource group**, and **Location** are locked and will be the same as they are for the virtual network gateway:

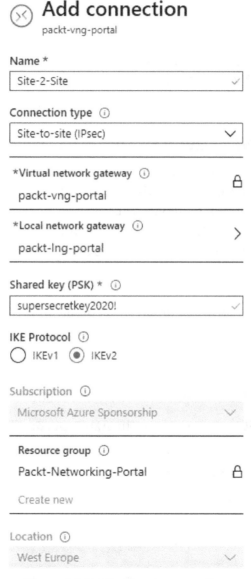

**Figure 8.4: Adding a new connection**

6.  Finally, we select **Create** and the deployment will start.

### How it works...

Using the virtual network gateway, we set up the Azure side of the IPsec tunnel. The local network gateway provides information on the local network, defining the local side of the tunnel with the public IP address and local subnet information. This way, Azure's side of the tunnel has all the relevant information that's needed to form a successful connection with an on-premises network. However, this completes only half of the work, as the opposite side of the connection must be configured as well. This part of the work really depends on the VPN device that's used locally, and each device has unique configuration steps. After both sides of the tunnel are configured, the result is a secure and encrypted VPN connection between networks.

Let's take a look at how to configure our local VPN device.

# Downloading the VPN device configuration from Azure

After creating the Azure side of the Site-to-Site connection, we still need to configure the local VPN device. The configuration depends upon the vendor and the device type. You can see all the supported devices at https://docs.microsoft.com/azure/ vpn-gateway/vpn-gateway-about-vpn-devices. In some cases, there is an option to download configuration for a VPN device directly from the Azure portal.

### Getting ready

Before you start, open the browser and go to the Azure portal at https://portal.azure. com.

### How to do it...

To download the VPN device configuration, we must follow these steps:

1.  Locate the **Site-2-Site** connection in the Azure portal. The **Overview** pane will be opened by default.

2.  Select the **Download configuration** option from the top of the pane:

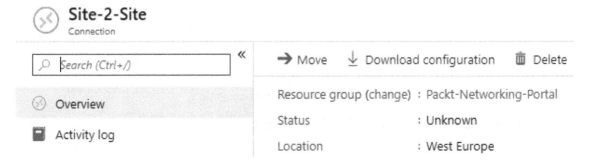

Figure 8.5: Site-2-Site connection overview in the Azure portal

3. A new pane will open, and you will see that all the options in the pane are predefined:

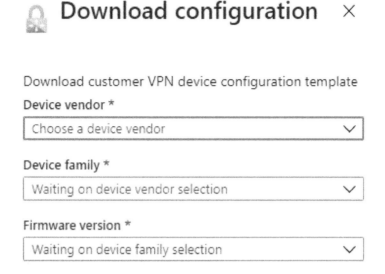

Figure 8.6: Choosing VPN device configuration

4. Select the relevant options for the **Device vendor**, **Device family**, and **Firmware version** fields. Note that only some options are available, and not all the supported devices have these options. After all of these options have been selected, download the configuration file. The sample file (`Site-2-Site.txt` in the `Chapter 8` folder) can be found in the GitHub repository associated with this book:

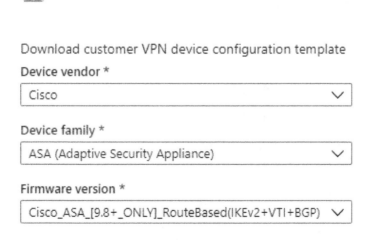

Figure 8.7: Downloading the configuration file

5. After using the configuration file for the local VPN device, both sides of the IPsec tunnel are configured. The **Status** value under the **Site-2-Site** connection will change to **Connected**:

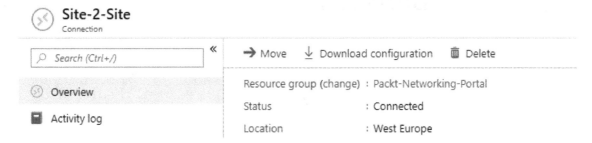

Figure 8.8: Checking the status of Site-2-Site connection

Now, let's have a look at how this connection functions in detail.

## How it works...

After we set up the Azure side of the IPsec tunnel, we need to configure the other side, as well as the local VPN device. The steps and configuration are different for each device. In some cases, we can download the configuration file directly from the Azure portal. After the VPN device has been configured, everything is set up, and we can use the tunnel for secure communication between the local network and the VNet.

# Creating a Point-to-Site connection

Accessing resources in a secure way is important, and this must be performed securely. It's not always possible to perform this using a Site-to-Site connection, especially when we have to perform something out of work hours. In this case, we can use Point-to-Site to create a secure connection that can be established from anywhere.

## Getting ready

To create a Point-to-Site connection, we'll need to generate a certificate that will be used for the connection. To create a certificate, we must follow these steps:

1. Execute the following PowerShell script to generate a certificate:

```
$cert = New-SelfSignedCertificate -Type Custom '

-KeySpec Signature '

-Subject "CN=P2SRootCert" '

-KeyExportPolicy Exportable '

-HashAlgorithm sha256 -KeyLength 2048 '

-CertStoreLocation "Cert:\CurrentUser\My" '
```

```
-KeyUsageProperty Sign '

-KeyUsage CertSign


New-SelfSignedCertificate -Type Custom '

-DnsName P2SChildCert '

-KeySpec Signature '

-Subject "CN=P2SChildCert" '

-KeyExportPolicy Exportable '

-HashAlgorithm sha256 -KeyLength 2048 '

-CertStoreLocation "Cert:\CurrentUser\My" '

-Signer $cert '

-TextExtension @("2.5.29.37={text}1.3.6.1.5.5.7.3.2")
```

2. Next, we need to export the certificate. Open **certmgr**, go to **Personal>Certificates**, select **P2SRootCert**, and then choose the **Export...** option:

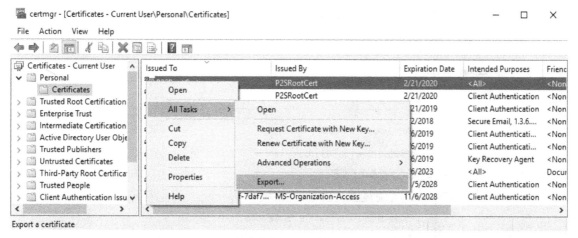

Figure 8.9: Exporting the certificate using certmgr

3. This will start the **Certificate Export Wizard**. Click **Next**.

4. Select the **No, do not export the private key** option and    click **Next**:

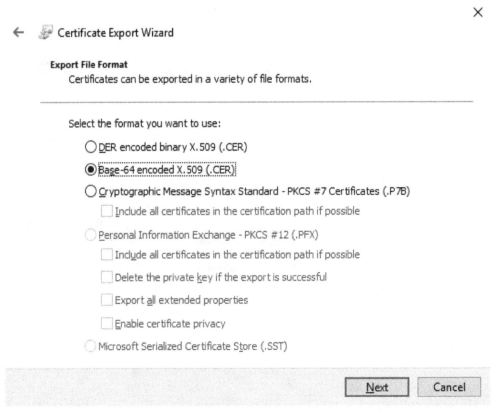

Figure 8.10: Certificate Export Wizard

5. Select the **Base-64 encoded X.509 (.CER)** format and click **Next**:

Figure 8.11: Selecting the export format

6. Select the location where you want to save the certificate and click **Next**.

7. Finally, we have the option to review all the information. After clicking **Finish**, the export will be complete:

Figure 8.12: Completing the Certificate Export Wizard

Now, let's look at the steps to create a Point-to-Site connection.

## How to do it...

To create a Point-to-Site connection, we need to do the following:

1. In the Azure portal, locate the virtual network gateway and **User VPN configuration**. Select **Configure now**:

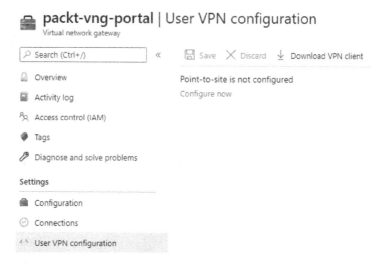

Figure 8.13: Configuring Point-to-Site connection

2. We need to define the **Address pool**. The address pool here cannot overlap with the address pool of the VNet associated with the virtual network gateway:

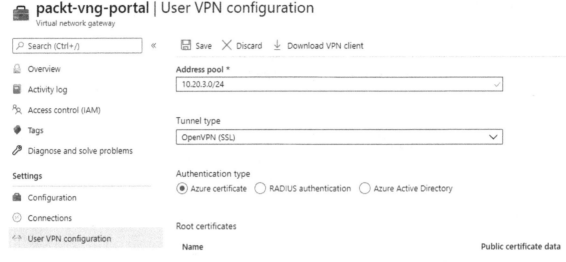

Figure 8.14: Adding the address pool

3. Next, we need to select a **Tunnel type** option from the list of predefined options. In this recipe, we'll select **OpenVPN (SSL)**, but any option is valid:

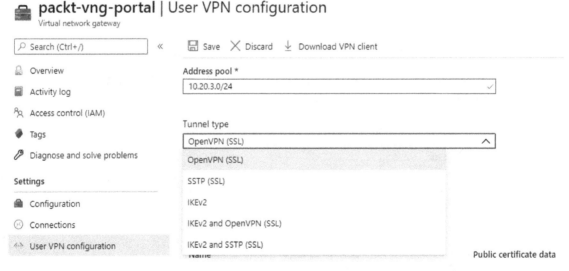

Figure 8.15: Selecting Tunnel type from the drop-down menu

4. Locate the exported certificate (from the *Getting ready* section) and open it in Notepad (or any other text editor). Select the value of the certificate and copy this value as follows:

Figure 8.16: Opening the certificate in Notepad

5. In the Azure portal, we need to define the root certificate. Enter the name of the certificate and then paste the value of the certificate (from the previous step) into the **Public certificate data** field:

Figure 8.17: Defining the root certificate

6. After clicking **Save** for the Point-to-Site configuration, a new option will become available: **Download VPN client**. We can download the configuration and start using this connection:

**Figure 8.18: Download the configuration**

Now, let's have a look at how it works.

## How it works...

Point-to-Site allows us to access Azure VNets in a secure way. Access to a Site-to-Site connection is restricted to our local network, but Point-to-Site allows us to connect from anywhere. Certificate-based authentication is used, which uses the same certificate on both the server (Azure) and the client (the VPN client) to verify the connection and permit access. This allows us to access Azure VNets from anywhere and at any time. This type of connection is usually used for management and maintenance tasks, as it's an on-demand connection. If a constant connection is needed, you need to consider a Site-to-Site connection.

# Creating a VNet-to-VNet connection

Similar to the need to connect Azure VNets to resources on a local network, we may have the need to connect to resources in another Azure VNet. In such cases, we can create a VNet-to-VNet connection that will allow us to use services and endpoints in another VNet. This process is very similar to creating a Site-to-Site connection; the difference is that we don't require a local network gateway. Instead, we use two virtual network gateways, one for each VNet.

## Getting ready

Before you start, open your browser and go to the Azure portal at https://portal.azure.com.

## How to do it...

To create a VNet-to-VNet connection, we must follow these steps:

1. In the Azure portal, locate one of the virtual network gateways (associated with one of the VNets you are trying to connect to).

2. In the **Virtual network gateway** pane, select **Connections** and select **Add** to add a new connection:

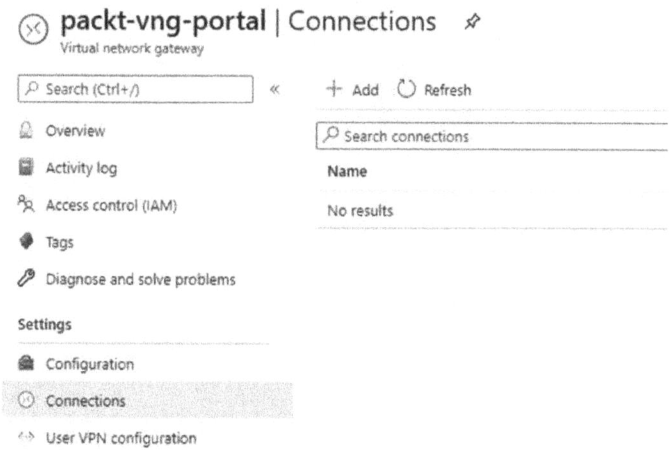

**packt-vng-portal** | Connections

Virtual network gateway

Search (Ctrl+/) «    + Add  ↻ Refresh

Overview                      Search connections

Activity log                  **Name**

Access control (IAM)          No results

Tags

Diagnose and solve problems

**Settings**

Configuration

Connections

User VPN configuration

Figure 8.19: Adding a new connection

3.  In the new pane, enter a **Name** value for the new connection and select **VNet-to-VNet** under **Connection type**:

**Add connection**

packt-vng-portal

Name *

VNet-2-VNet

Connection type ⓘ

VNet-to-VNet

*First virtual network gateway ⓘ

packt-vng-portal

*Second virtual network gateway ⓘ

Choose another virtual network gateway

Shared key (PSK) * ⓘ

IKE Protocol ⓘ

◯ IKEv1  ⦿ IKEv2

Subscription ⓘ

Microsoft Azure Sponsorship

Resource group ⓘ

Packt-Networking-Portal

Create new

Location ⓘ

West Europe

Figure 8.20: Configuring the new connection

4. The first virtual network gateway will be automatically highlighted. We need to select the second virtual network gateway:

**Figure 8.21: Choosing the virtual network gateway**

5. We need to provide a shared key for our connection before we select **Create** and start the deployment. Note that **Subscription**, **Resource group**, and **Location** are locked and that the values for the first virtual network gateway are used here:

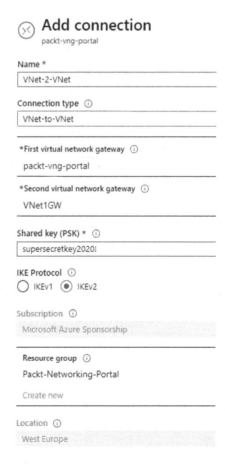

**Figure 8.22: Providing a shared key for the connection**

6. The deployment of VNet-to-VNet doesn't take long and should be done in a few minutes. However, it takes some time to establish connections, so you may see the status **Unknown** for up to 15 minutes before it changes to **Connected**:

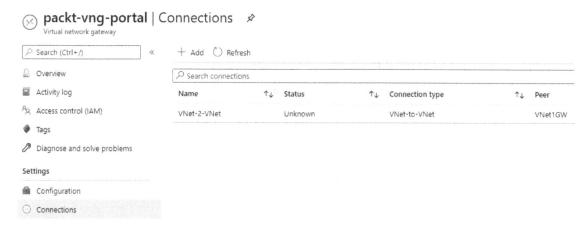

Figure 8.23: Deployment status of VNet-to-VNet

Now, let's have a look at its functioning in detail.

## How it works...

A VNet-to-VNet connection works very similarly to a Site-to-Site connection. The difference is that Azure uses a local network gateway for information on the local network. In this case, we don't need this information; we use two virtual network gateways to connect. Each virtual network gateway provides network information for the VNet that it's associated with. This results in secure, encrypted VPN connections between two Azure VNets that can be used to establish connections between Azure resources on both VNets.

Now, let's learn about using network peering to connect VNets.

## Connecting VNets using network peering

Another way to connect two Azure VNets is to use **network peering**. This approach doesn't require the use of a virtual network gateway, so it's more economical to use it if the only requirement is to establish a connection between Azure VNets. Network peering uses the Microsoft backbone infrastructure to establish a connection between two VNets, and traffic is routed through private IP addresses only. However, this traffic is not encrypted; it's private traffic that stays on the Microsoft network, similar to what happens to traffic on the same Azure VNet.

## Getting ready

Before you start, open your browser and go to the Azure portal at https://portal.azure.com.

## How to do it...

To create network peering, we must take the following steps:

1. In the Azure portal, locate one of the VNets that you want to connect to.

2. In the **Virtual network** pane, select the **Peerings** option, and select **Add** to add a new connection:

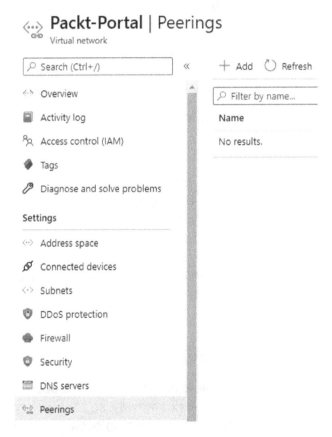

Figure 8.24: Adding a new network peering connection

3. In the new pane, we must enter the name of the connection, select a **Virtual network deployment model** option (**Resource manager** or **Classic**), and select the VNet we are connecting to. This information can be provided either by providing a resource ID or by selecting **Subscription** and **Virtual network** options from the drop-down menu. There are some additional configurations that are optional but provide us with better traffic control:

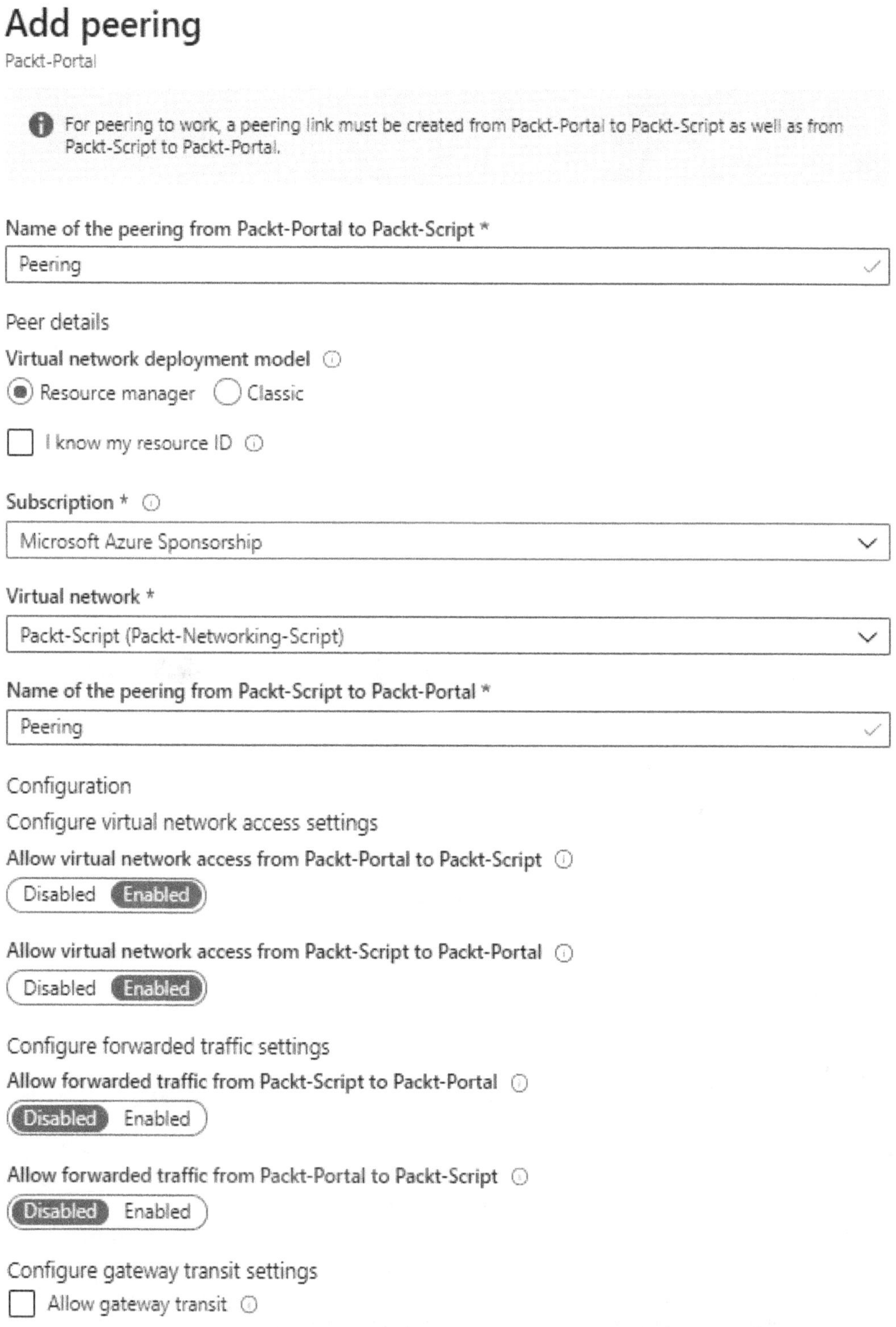

# Add peering

Packt-Portal

ⓘ For peering to work, a peering link must be created from Packt-Portal to Packt-Script as well as from Packt-Script to Packt-Portal.

**Name of the peering from Packt-Portal to Packt-Script \***

| Peering ✓ |
|---|

Peer details

**Virtual network deployment model** ⓘ

⦿ Resource manager    ◯ Classic

☐ I know my resource ID ⓘ

**Subscription \*** ⓘ

| Microsoft Azure Sponsorship ⌄ |
|---|

**Virtual network \***

| Packt-Script (Packt-Networking-Script) ⌄ |
|---|

**Name of the peering from Packt-Script to Packt-Portal \***

| Peering ✓ |
|---|

Configuration

Configure virtual network access settings

**Allow virtual network access from Packt-Portal to Packt-Script** ⓘ

( Disabled **Enabled** )

**Allow virtual network access from Packt-Script to Packt-Portal** ⓘ

( Disabled **Enabled** )

Configure forwarded traffic settings

**Allow forwarded traffic from Packt-Script to Packt-Portal** ⓘ

( **Disabled** Enabled )

**Allow forwarded traffic from Packt-Portal to Packt-Script** ⓘ

( **Disabled** Enabled )

Configure gateway transit settings

☐ Allow gateway transit ⓘ

**Figure 8.25: Configuring peer details for a new connection**

4. After a connection is created, we can see the information and the status for peering. We can also change the **Configuration** options at any time:

# Peering

Packt-Portal

💾 Save   ✕ Discard   🗑 Delete

---

Name of the peering from Packt-Portal to Packt-Script

Peering

Peering status

Connected

Provisioning state

Succeeded

Peer details

Address space

10.11.0.0/16

Remote Vnet Id

/subscriptions/cb638267-a366-463c-bfe5-7a49311c27a8/resourceGroups/Packt-Networking-Scri...  🗇

Virtual network

Packt-Script

Configuration

Configure virtual network access settings

Allow virtual network access from Packt-Portal to Packt-Script ⓘ

( Disabled  **Enabled** )

Configure forwarded traffic settings

Allow forwarded traffic from Packt-Script to Packt-Portal ⓘ

( **Disabled**  Enabled )

Configure gateway transit settings

☐  Allow gateway transit ⓘ

Configure Remote Gateways settings

▨  Use remote gateways ⓘ

**Figure 8.26: Reviewing the peering information and status for a new connection**

Now, let's have a look at its inner workings in detail.

## How it works...

Network peering allows us to establish a connection between two Azure VNets in the same Azure tenant. Peering uses a Microsoft backbone network to route private traffic between resources on the same network, using private IP addresses only. There is no need for virtual network gateways (which create additional cost), as a virtual "remote gateway" is created to establish a connection. The downside of this approach is that the same VNet can't use peering and a virtual network gateway at the same time. If there is a need to connect VNet to both the local network and another VNet, we must take a different approach and use a virtual network gateway, which will allow us to create a Site-to-Site connection with a local network and a VNet-to-VNet connection with another VNet.

When it comes to network access settings, we have multiple options to control network traffic flow. For example, we can say that traffic is allowed from VNet A to VNet B, but denied from VNet B to VNet A. Of course, we can set it the other way around or make it bidirectional.

We can also control transit traffic when an additional network is in the mix. If VNet A is connected to VNet B, and additionally VNet A is connected to VNet C, we can control whether traffic is allowed between VNet B and VNet C as transit traffic through VNet A.

However, this only works if transit is not made via peering. If all networks are Azure VNets, and VNet A connected to VNet B via peering, and VNet B connected to VNet C via peering, the connection between VNet A and VNet C would not be possible via transit between VNets. This is because peering is a non-transitive relationship between two VNets. If VNet B is connected to VNet C via VNet-to-VNet (or to an on-premises network via Site-to-Site), transit would be possible between VNet A and VNet C over VNet B.

# 9

# Connecting to resources securely

Exposing management endpoints (RDP, SSH, HTTP, and others) over a public IP address is not a good idea. Any kind of management access should be controlled and allowed only over a secure connection. Usually, this is done by connecting to a private network (via S2S or P2S) and accessing resources over private IP addresses. In some situations, this is not easy to achieve. The cause of this can be insufficient local infrastructure, or in some cases, the scenario may be too complex. Fortunately, there are other ways to achieve the same goal. We can safely connect to our resources using Azure Bastion, Azure Virtual WAN, and Azure Private Link.

We will cover the following recipes in this chapter:

- Creating an Azure Bastion instance
- Connecting to a virtual machine with Azure Bastion
- Creating a virtual WAN
- Creating a hub (in Virtual WAN)
- Adding a Site-to-Site connection (in a virtual hub)
- Adding a virtual network connection (in a virtual hub)
- Creating a Private Link endpoint
- Creating a Private Link service

## Technical requirements

For this chapter, the following is required:

- An Azure subscription

## Creating an Azure Bastion instance

Azure Bastion allows us to connect securely to our Azure resources without additional infrastructure. All we need is a browser. It is essentially a PaaS service provisioned in our virtual network that provides a secure RDP/SSH connection to Azure Virtual Machines. The connection is made directly from the Azure portal over **Transport Layer Security (TLS)**.

## Getting ready

Before we can create an Azure Bastion instance, we must prepare the subnet.

In order to create a new subnet for Azure Bastion, we must do the following:

1. Locate the virtual network that will be associated with our Azure Bastion instance.

2. Select the **Subnets** option under **Settings** and select the option to add a new subnet, as shown in *Figure 9.1*:

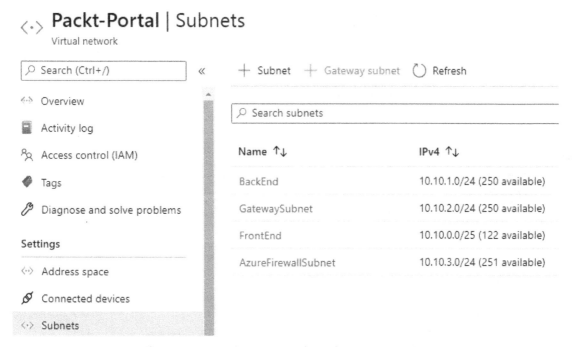

Figure 9.1: Creating a new subnet for Azure Bastion

3. In the new pane, we must fill in the **Name** and **Address range** fields. It's very important that the subnet is named **AzureBastionSubnet** and that the subnet uses a prefix of at least **/27** (this is a service requirement, and we will not be able to proceed otherwise). Options for **NAT gateway** and **Network security group (NSG)** can be added if needed (for example, a rule forcing traffic through **network address translation (NAT)**). The **Service endpoints** and **Subnet delegation** fields are not required, and as this subnet should be dedicated to Azure Bastion only, it is not recommended to use them:

Figure 9.2: Fill in Name and Address range for the subnet

## How to do it...

In order to create a new Azure Bastion instance, we must follow these steps:

1. In the Azure portal, select **Create a resource** and choose **Azure Bastion** under **Networking** (or search for `Azure Bastion` in the search bar).

2. In the new pane, we must provide information for the **Subscription**, **Resource group**, **Name**, and **Region** fields. Next, we must make a selection for **Virtual network** (only networks in the same region will be available) and **Subnet** (the one that we previously created) and provide information for **Public IP address** (select an existing one or create a new one):

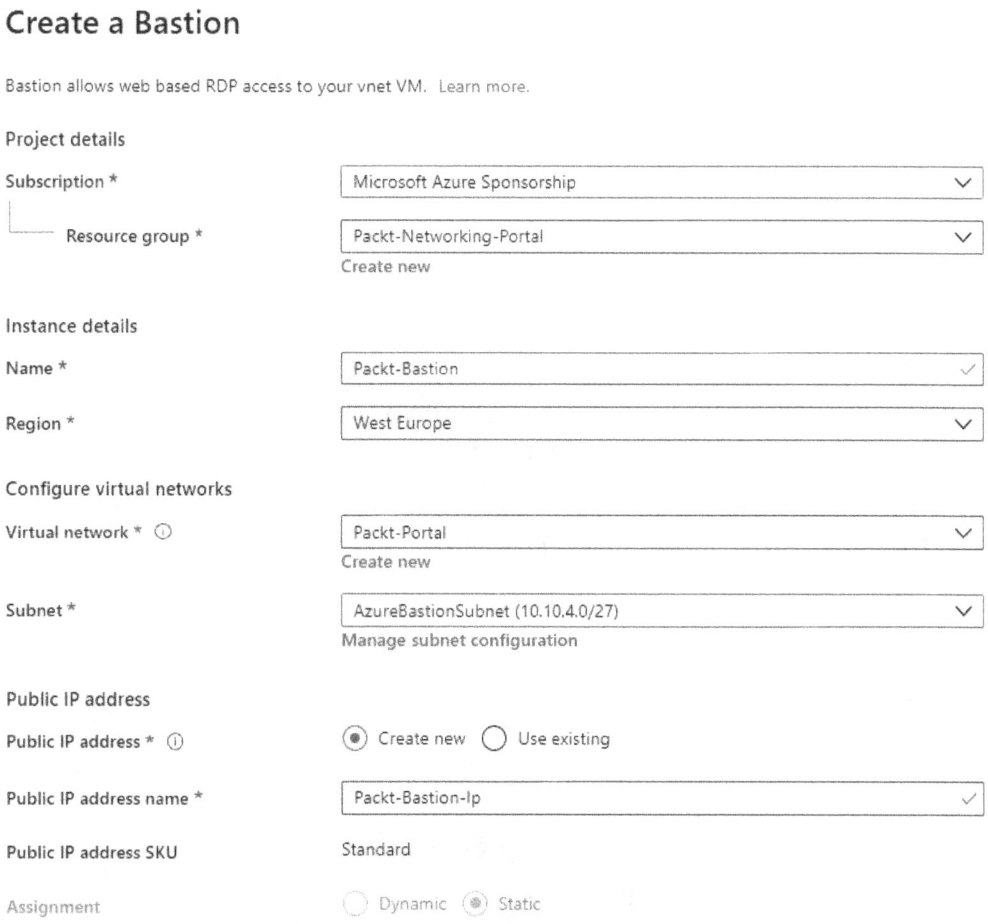

## Create a Bastion

Bastion allows web based RDP access to your vnet VM.  Learn more.

Project details

| Subscription * | Microsoft Azure Sponsorship |
|---|---|
| Resource group * | Packt-Networking-Portal |
| | Create new |

Instance details

| Name * | Packt-Bastion |
|---|---|
| Region * | West Europe |

Configure virtual networks

| Virtual network * ⓘ | Packt-Portal |
|---|---|
| | Create new |
| Subnet * | AzureBastionSubnet (10.10.4.0/27) |
| | Manage subnet configuration |

Public IP address

| Public IP address * ⓘ | ◉ Create new  ◯ Use existing |
|---|---|
| Public IP address name * | Packt-Bastion-Ip |
| Public IP address SKU | Standard |
| Assignment | ◯ Dynamic  ◉ Static |

Figure 9.3: Configuration details of a Bastion instance

## How it works...

Azure Bastion is provisioned inside our virtual network, which allows communication with all resources on that network. Using TLS, it provides a secure RDP and SSH connection to all resources on that network. The connection is made through a browser session, and no public IP address is required. This means that we don't need to expose any of the management ports over a public IP address.

After creating the Azure Bastion instance, let's move on to the next recipe, where we will learn how to connect to a virtual machine with Azure Bastion.

# Connecting to a virtual machine with Azure Bastion

With Azure Bastion, we can connect to a virtual machine through the browser without a public IP address and without exposing it publicly.

## Getting ready

Before you start, open the browser and go to the Azure portal via https://portal.azure. com.

## How to do it...

In order to connect to a virtual machine with Azure Bastion, we must follow these steps:

1.  In the Azure portal, find the virtual machine you want to connect to. The virtual machine needs to be on the same virtual network as Azure Bastion is deployed on.

2.  In the **Virtual machine** pane, select the **Connect** option under **Settings**. Select the **BASTION** tab, and on that tab, select **Use Bastion**:

Figure 9.4: Connecting to a virtual machine with Azure Bastion

3. Select the **Open in new window** option and fill in **Username** and **Password**:

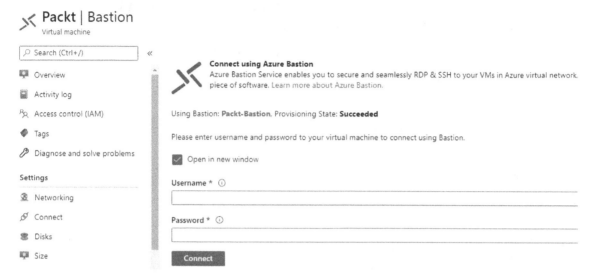

Figure 9.5: Adding a username and password for the virtual machine

The connection will open in a new window, allowing you to fully manage your virtual machine. The interface depends on the default management port, RDP or SSH.

## How it works...

Azure Bastion uses a subnet in the virtual network to connect to virtual machines in that specific network. It provides a safe connection over TLS and allows a connection to a virtual machine without exposing it over a public IP address.

In this recipe, we learned how to connect a virtual machine with Azure Bastion. In the next recipe, we'll learn how to create a virtual WAN.

# Creating a virtual WAN

In many situations, the network topology can get very complex. It can be difficult to keep track of all network connections, gateways, and peering processes. Azure Virtual WAN provides a single interface to manage all these points.

## Getting ready

Before you start, open the browser and go to the Azure portal via https://portal.azure.com.

## How to do it...

1.  In the Azure portal, select **Create a resource** and choose **Virtual WAN** under **Networking** (or search for `Virtual WAN` in the search bar).

2.  In the new pane, we must provide information for the **Subscription**, **Resource group**, **Resource group location**, **Name**, and **Type** fields:

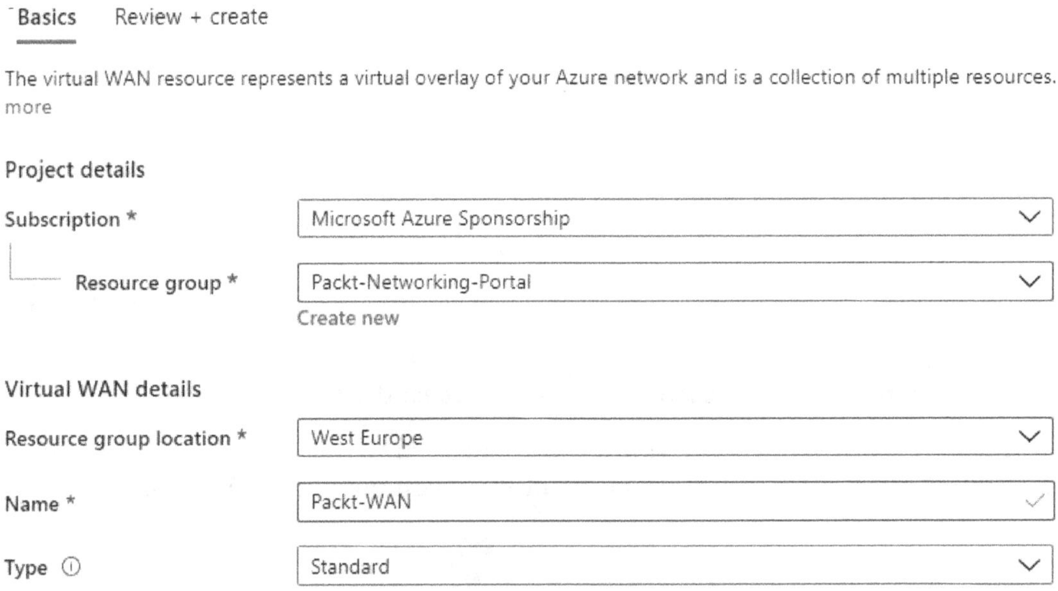

Basics    Review + create

The virtual WAN resource represents a virtual overlay of your Azure network and is a collection of multiple resources.  Learn more

Project details

Subscription *          | Microsoft Azure Sponsorship                     ∨ |

    Resource group *   | Packt-Networking-Portal                         ∨ |
                         Create new

Virtual WAN details

Resource group location * | West Europe                                    ∨ |

Name *                  | Packt-WAN                                       ✓ |

Type ⓘ                  | Standard                                        ∨ |

Figure 9.6: Information for the virtual WAN resource

Azure Virtual WAN is ready for deployment and it usually takes only a few minutes to complete.

## How it works...

Azure Virtual WAN brings multiple network services to a single point. From here, we can configure, control, and monitor connections such as Site-to-Site, Point-to-Site, ExpressRoute, or a connection between virtual networks. When we have multiple Site-to-Site connections or multiple virtual networks connected with peering, it can be hard to keep track of all these resources. Virtual WAN allows us to do that with a single service.

This is accomplished with hubs, and in the next recipe, we'll see how to set one up.

# Creating a hub (in Virtual WAN)

Hubs are used as regional connection points. They contain multiple service endpoints that enable connectivity between different networks and services. They're the core of networking for each region.

## Getting ready

Before you start, open the browser and go to the Azure portal via https://portal.azure.com.

## How to do it...

1. In the Azure portal, locate the previously created virtual WAN.

2. In the **Virtual WAN** pane, select **Hubs** under the **Connectivity** section. Select the option to add a new hub:

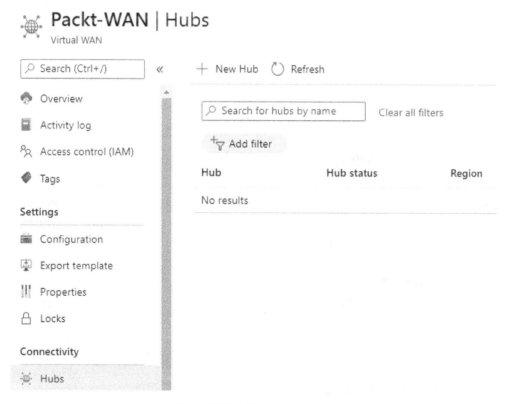

Figure 9.7: Adding a new hub

3.  In the new pane, we need to provide information in the **Region**, **Name** (for the new hub), and **Hub private address space** fields. **Subscription** and **Resource group** are grayed out as they use the same options as Virtual WAN:

## Create virtual hub

Basics   Site to site   Point to site   ExpressRoute   Tags   Review + create

A virtual hub is a Microsoft-managed virtual network. The hub contains various service endpoints to enable connectivity from your on-premises network (vpnsite). The hub is the core of your network in a region. There can only be one hub per Azure region. When you create a hub using Azure portal, it creates a virtual hub VNet and a virtual hub vpngateway.   Learn more

**Project details**

The hub will be created under the same subscription and resource group as the vWAN.

| | |
|---|---|
| Subscription | Microsoft Azure Sponsorship |
| └─ Resource group | Packt-Networking-Portal |

**Virtual Hub Details**

| | |
|---|---|
| Region * | West Europe |
| Name * | Hub1 |
| Hub private address space * ⓘ | 192.168.0.0/16 |

**Figure 9.8: Information for the new virtual hub**

4.  The next three steps are optional, and we can choose any or all of them. The first step is to configure a Site-to-Site gateway. If we enable this option, we need to select an option for **Gateway scale units** (or SKU). An autonomous system number (**AS Number**) is provided to be used if needed (for VPN configuration later):

## Create virtual hub

Basics   **Site to site**   Point to site   ExpressRoute   Tags   Review + create

You will need to enable Site to site (VPN gateway) before connecting to VPN sites. You can do this after hub creation, but doing it now will save time and reduce the risk of service interruptions later.   Learn more

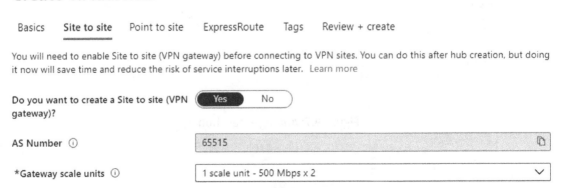

| | |
|---|---|
| Do you want to create a Site to site (VPN gateway)? | Yes    No |
| AS Number ⓘ | 65515 |
| *Gateway scale units ⓘ | 1 scale unit - 500 Mbps x 2 |

**Figure 9.9: Configuring a Site-to-Site gateway**

5. The next optional setting is **Point to site**. If we choose to enable it, we need to select an option for **Gateway scale units** and **Point to site configuration**. Click on **Create new** to add a new configuration:

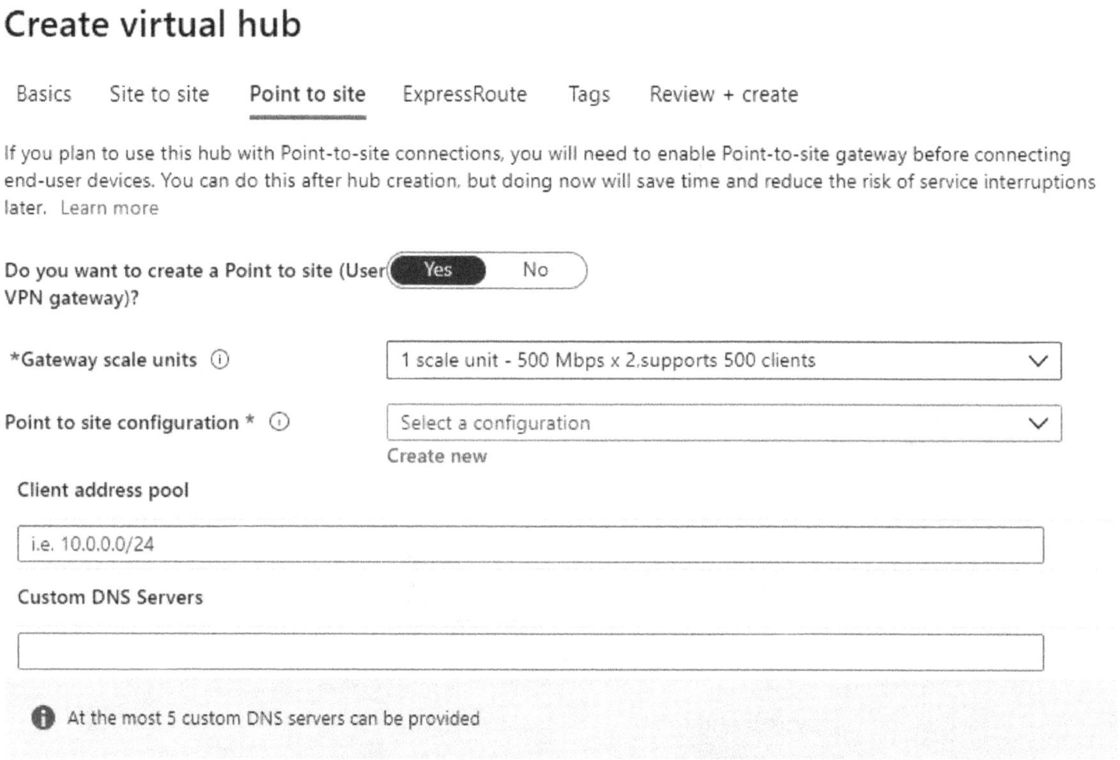

# Create virtual hub

Basics    Site to site    **Point to site**    ExpressRoute    Tags    Review + create

If you plan to use this hub with Point-to-site connections, you will need to enable Point-to-site gateway before connecting end-user devices. You can do this after hub creation, but doing now will save time and reduce the risk of service interruptions later. Learn more

Do you want to create a Point to site (User VPN gateway)?    [ Yes    No ]

*Gateway scale units ⓘ    [ 1 scale unit - 500 Mbps x 2.supports 500 clients    ⌄ ]

Point to site configuration * ⓘ    [ Select a configuration    ⌄ ]
Create new

Client address pool

[ i.e. 10.0.0.0/24 ]

Custom DNS Servers

[                                                          ]

ⓘ At the most 5 custom DNS servers can be provided

Figure 9.10: Configuring a Point-to-Site gateway

6.  In the new pane, we need to provide information for **Configuration name**, **Tunnel type**, and **Authentication method**. If **Azure certificate** is used, we need to provide certificate information (for more information about certificates, see the *Creating a Point-to-Site connection* recipe from *Chapter 8, Creating hybrid connections*):

**Figure 9.11: Creating a new VPN configuration**

7.  After the Point-to-Site configuration is added, we are returned to the previous pane. We need to fill in the **Client address pool** field, and optionally **Custom DNS Servers**:

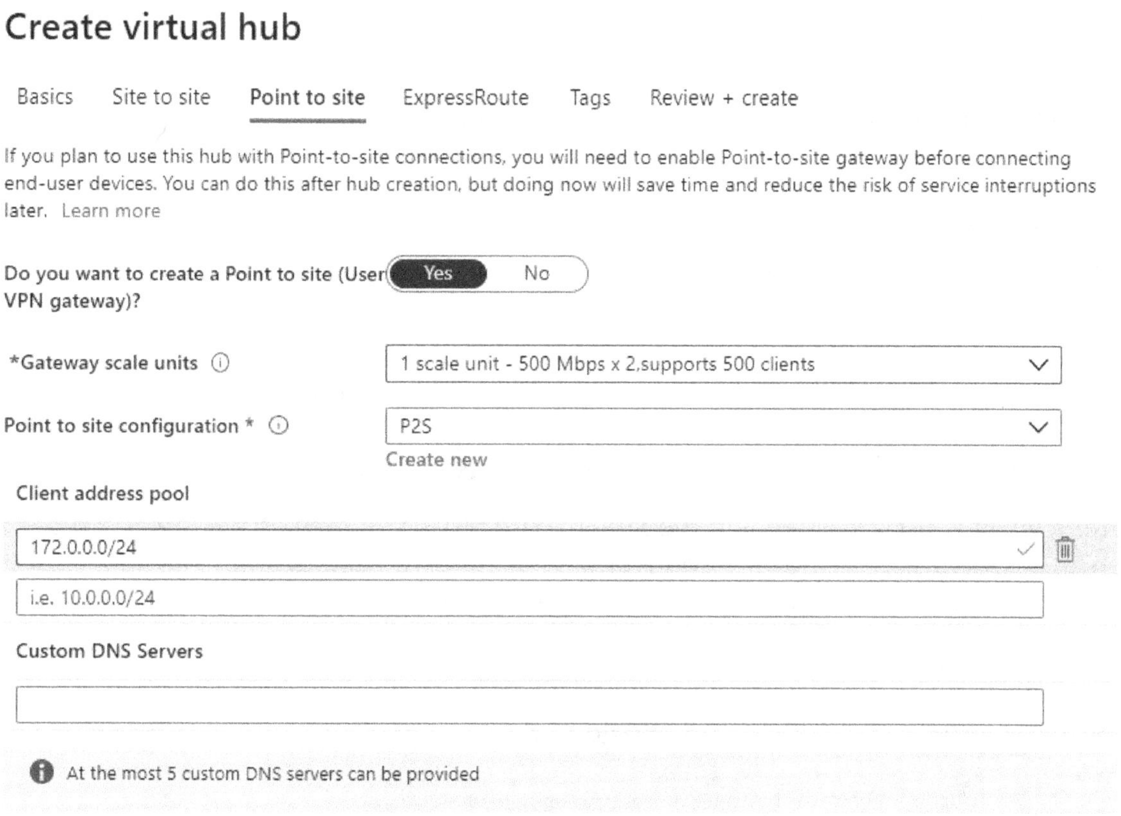

# Create virtual hub

Basics    Site to site    **Point to site**    ExpressRoute    Tags    Review + create

If you plan to use this hub with Point-to-site connections, you will need to enable Point-to-site gateway before connecting end-user devices. You can do this after hub creation, but doing now will save time and reduce the risk of service interruptions later.  Learn more

Do you want to create a Point to site (User VPN gateway)?     **Yes**    No

*Gateway scale units  ⓘ

> 1 scale unit - 500 Mbps x 2,supports 500 clients    ⌄

Point to site configuration *  ⓘ

> P2S    ⌄
>
> Create new

Client address pool

> 172.0.0.0/24    ✓    🗑

> i.e. 10.0.0.0/24

Custom DNS Servers

> 

ⓘ At the most 5 custom DNS servers can be provided

Figure 9.12: Adding Client address pool and Custom DNS Servers information

8.  The third optional setting is **ExpressRoute**. If we choose to enable it, we need to select an option for **Gateway scale units**:

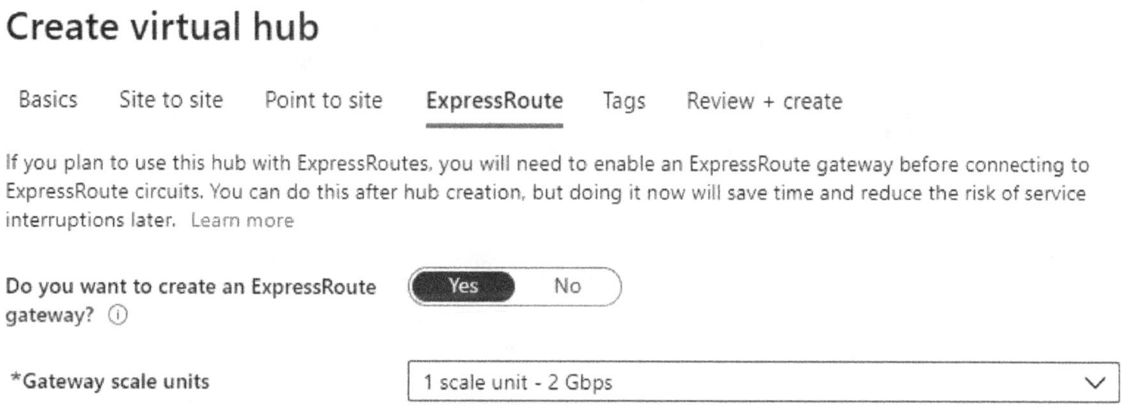

# Create virtual hub

Basics    Site to site    Point to site    **ExpressRoute**    Tags    Review + create

If you plan to use this hub with ExpressRoutes, you will need to enable an ExpressRoute gateway before connecting to ExpressRoute circuits. You can do this after hub creation, but doing it now will save time and reduce the risk of service interruptions later.  Learn more

Do you want to create an ExpressRoute gateway? ⓘ     **Yes**    No

*Gateway scale units

> 1 scale unit - 2 Gbps    ⌄

Figure 9.13: Configuring ExpressRoute

9.  We can optionally add tags, and then proceed with the creation of the virtual hub. It can take up to 30 minutes to complete deployment.

## How it works...

Virtual hubs represent control points inside a region. From there, we can define all connections to virtual networks inside the region. This applies to Site-to-Site, Point-to-Site, and ExpressRoute. Each section is optional, and we can create a hub without any configurations for connection types. If we choose to create them at this point, we need to provide an SKU for each type. A Point-to-Site connection also requires the user's VPN configuration to be provided. Each connection type can be added at a later time as well.

In this recipe, we learned how to create a virtual hub. Let's move on to the next recipe and learn how to add a Site-to-Site connection in a virtual hub.

# Adding a Site-to-Site connection (in a virtual hub)

After a virtual hub is created and the Site-to-Site SKU is defined inside the hub, we can proceed to create a Site-to-Site connection. For this, we need to apply the appropriate connection settings and provide configuration details.

## Getting ready

Before you start, open the browser and go to the Azure portal via https://portal.azure.com.

## How to do it...

In order to create a Site-to-Site connection in a virtual hub (under a virtual WAN), we must take the following steps:

1. Find the virtual WAN and locate the previously created virtual hub under **Hubs** in the **Connectivity** section. Select that hub:

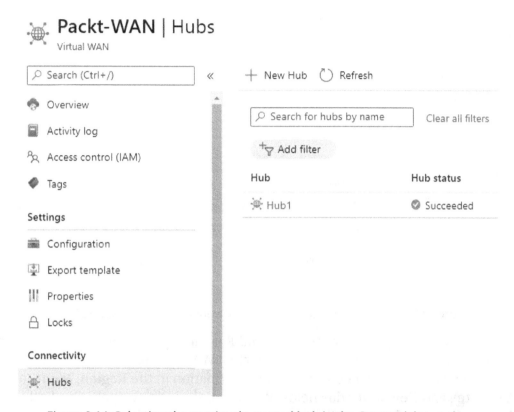

Figure 9.14: Selecting the previously created hub in the Connectivity section

2.  In the **Virtual HUB** pane, go to the **VPN (Site to site)** settings under **Connectivity**. Select the **Create new VPN site** option:

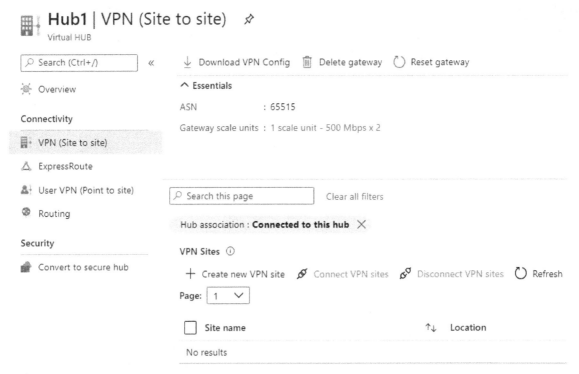

Figure 9.15: Selecting the Create new VPN site option in the Virtual HUB pane

3.  A new pane will appear. **Subscription** and **Resource group** are grayed out, as the VPN site is a child resource under the virtual WAN and must use the same options as the virtual WAN. We need to provide information in the **Region**, **Name** (of the VPN site), and **Device vendor** fields. We have the option to enable or disable **Border Gateway Protocol (BGP)**. If BGP is not configured, we need to provide at least one private address space. We also need to define a hub (or more of them) that will be used in the connection:

## Create VPN site

**Project details**

Subscription                                    Microsoft Azure Sponsorship          ∨

     Resource group                          Packt-Networking-Portal            ∨

**Instance details**

Region *                                        West Europe                         ∨

Name *                                          VPN1                                ✓

Device vendor *                                 Palo Alto                           ✓

Border Gateway Protocol                         ( **Enable**   Disable )

Private address space

  At least one address space is required if BGP isn't configured

[                                                                                  ]

Connect to

Hubs ⓘ

[ Hub1                                                                    ∨ ] 🗑

[                                                                         ∨ ]

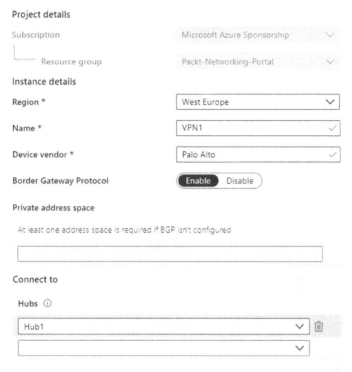

Figure 9.16: Creating a VPN site

4. In the **Links** section of the VPN site, we need to provide information for **Link name**, **Provider name**, **Speed** (in Mbps), **IP address / FQDN** (of the VPN device we want to connect to), **BGP address**, and **ASN** as shown in *Figure 9.17*:

Basics    **Links**    Review + create

Link Details ⓘ              + Add

☐

Link name * ⓘ    [ Link1            ✓ ]    Provider name * ⓘ [ Logosoft        ✓ ]

Speed * ⓘ        [ 100              ✓ ]    IP address / FQDN * [ 217.75.192.10   ✓ ]
                                          ⓘ

BGP address *    [ 192.168.150.5    ✓ ]    ASN *             [ 64512           ✓ ]

Figure 9.17: Providing link details in the Links pane

5. After the VPN site is created, we can download the VPN configuration for the VPN device. After the VPN device is configured, we can select the VPN site and initiate the connection with the **Connect VPN sites** option:

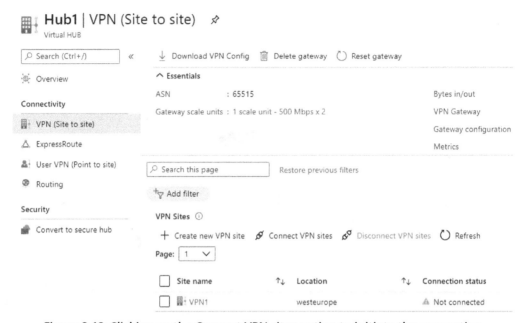

Figure 9.18: Clicking on the Connect VPN sites option to initiate the connection

6. This will open a new pane. We must provide information for **Pre-shared key (PSK)**, **Protocol** and **IPSec**, and choose options for **Propagate Default Route** and **Use policy based traffic selector**:

Figure 9.19: Providing information in the Connect sites pane

## How it works...

Adding a Site-to-Site connection to our virtual hub allows us to connect to a virtual hub in a specific region from our on-premises network (or other networks using **Virtual appliance**). To do so, we must provide information about the VPN connection in the virtual hub and configure the VPN device that will be used to connect.

However, this allows us only to connect to the hub. We need to connect virtual networks in order to access Azure resources. In the next recipe, we'll see how to add a virtual network connection to the virtual hub.

# Adding a virtual network connection (in a virtual hub)

A virtual hub represents a central point in an Azure region. But to actually use this point, we need to connect virtual networks to a virtual hub. Then, we can use the virtual hub as intended.

## Getting ready

Before you start, open the browser and go to the Azure portal via https://portal.azure.com.

## How to do it...

In order to add a virtual network connection in a virtual hub (under a virtual WAN), we must take the following steps:

1. Find the virtual WAN and locate the previously created virtual hub under **Virtual network connections** in the **Connectivity** section. Select the **Add connection** option:

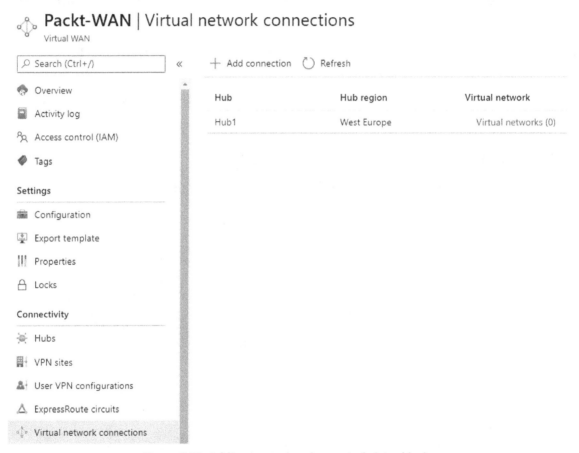

**Figure 9.20: Adding a previously created virtual hub**

2. In the new pane, we need to provide information in the **Connection name**, **Hubs**, **Subscription**, **Resource group**, and **Virtual network** fields. Next, we need to provide **Routing configuration** information. We can select **Yes** for **Propagate to none**. If we select **No**, we need to provide information for **Associate Route Table**, **Propagate to Route Tables**, and **Propagate to labels**. **Static routes** is an optional setting:

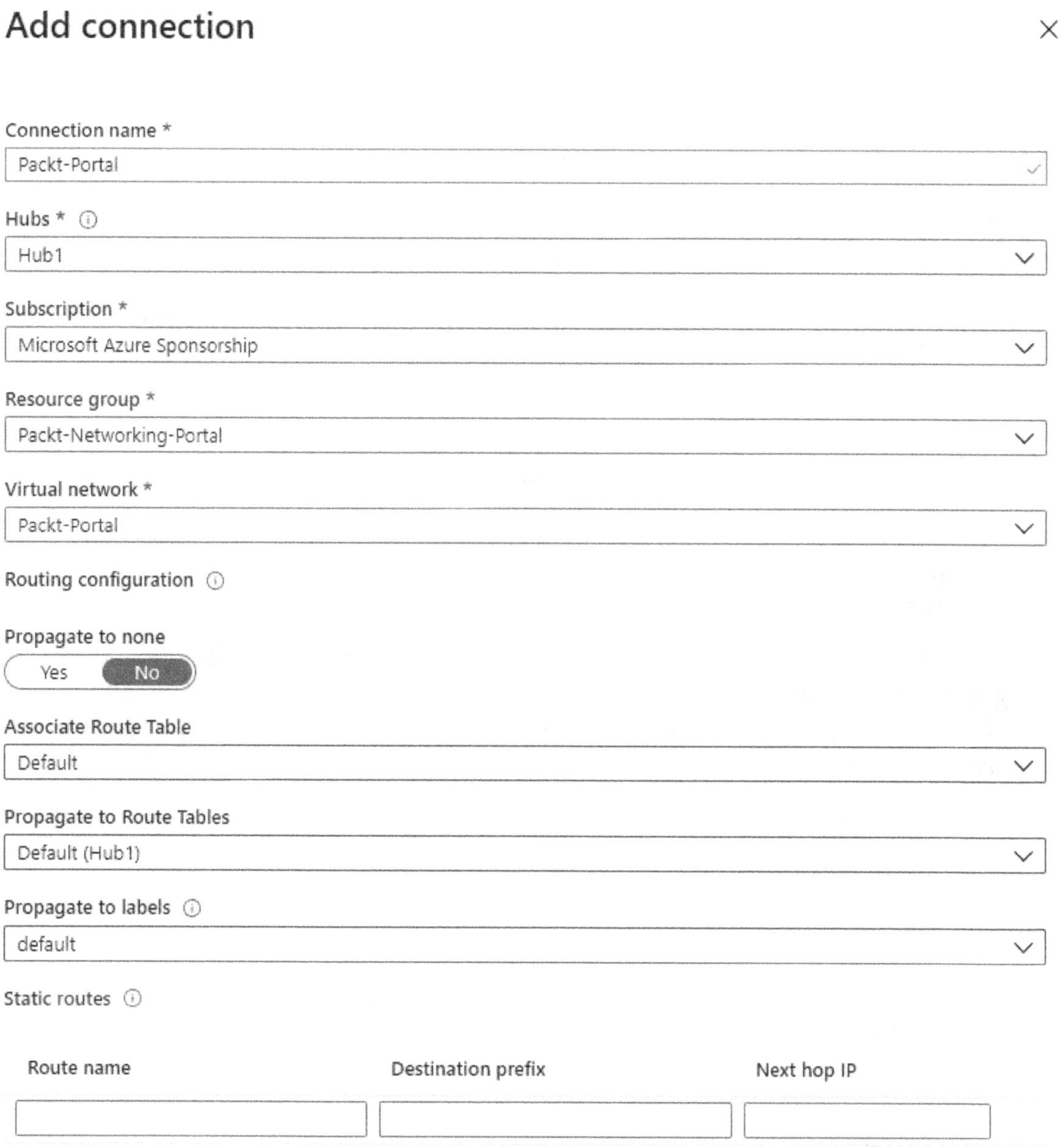

## Add connection                                        ✕

Connection name *

| Packt-Portal | ✓ |

Hubs *  ⓘ

| Hub1 | ∨ |

Subscription *

| Microsoft Azure Sponsorship | ∨ |

Resource group *

| Packt-Networking-Portal | ∨ |

Virtual network *

| Packt-Portal | ∨ |

Routing configuration  ⓘ

Propagate to none

( Yes  **No** )

Associate Route Table

| Default | ∨ |

Propagate to Route Tables

| Default (Hub1) | ∨ |

Propagate to labels  ⓘ

| default | ∨ |

Static routes  ⓘ

| Route name | Destination prefix | Next hop IP |
|---|---|---|
|  |  |  |

Figure 9.21: Configuring the virtual hub details

## How it works...

Connecting a virtual network to a virtual hub will allow us to access resources when connected to the same hub. A connection can be made over a Site-to-Site connection, a Point-to-Site connection, or from another virtual network (connected to the same hub). When creating a connection, we need to provide routing and propagation rules in order to define the network flow. We can also define a static route. A static route will force all traffic to go through a single IP address, usually through a firewall or network virtual appliance.

Let's move on to the next recipe and learn how to create a Private Link endpoint.

# Creating a Private Link endpoint

Private Link allows us to connect to PaaS services over a secure network. As these services are usually exposed over the internet, this gives us a more secure method of access. There are two components available to make a secure connection—a Private Link endpoint and a Private Link service. Let's start by creating a Private Link endpoint first.

## Getting ready

We need to create a service that will be associated with the Private Link endpoint:

1.  Open the browser and go to the Azure portal via https://portal.azure.com. Select the option to create a new service. Search for **SQL Server** (logical server) and select the **Create new** option.

2.  In the new pane, we must provide information in the **Subscription**, **Resource group**, **Server name** (must be a unique FQDN), and **Location** fields. Finally, we must provide credentials for the administrator login before selecting **Review + create**:

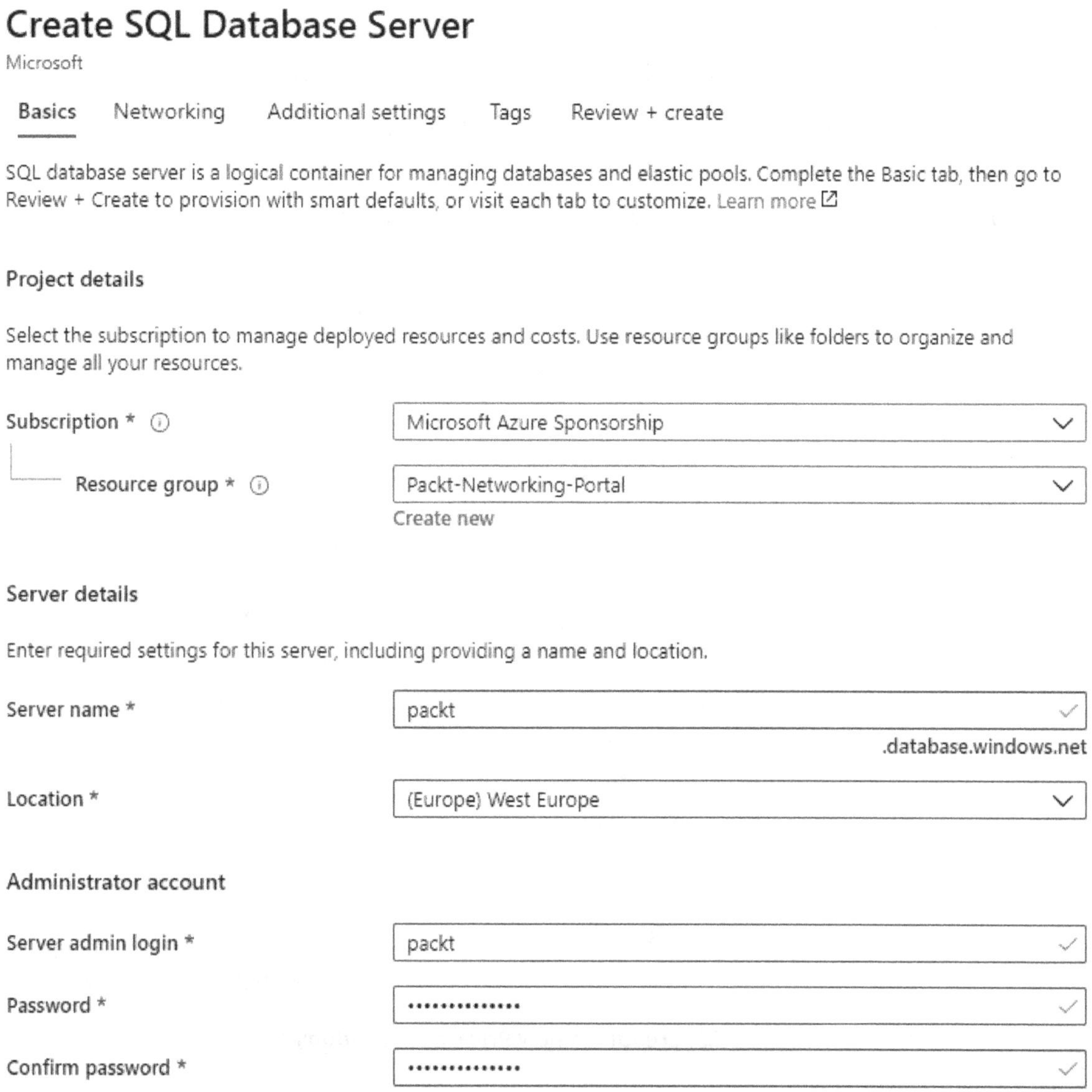

# Create SQL Database Server

Microsoft

**Basics**   Networking   Additional settings   Tags   Review + create

SQL database server is a logical container for managing databases and elastic pools. Complete the Basic tab, then go to Review + Create to provision with smart defaults, or visit each tab to customize. Learn more ☐

## Project details

Select the subscription to manage deployed resources and costs. Use resource groups like folders to organize and manage all your resources.

Subscription * ⓘ

| Microsoft Azure Sponsorship | ⌄ |

　　　　Resource group * ⓘ

| Packt-Networking-Portal | ⌄ |

Create new

## Server details

Enter required settings for this server, including providing a name and location.

Server name *

| packt | ✓ |

.database.windows.net

Location *

| (Europe) West Europe | ⌄ |

## Administrator account

Server admin login *

| packt | ✓ |

Password *

| •••••••••••••• | ✓ |

Confirm password *

| •••••••••••••• | ✓ |

**Figure 9.22: Associating a new service with a Private Link endpoint**

## How to do it...

In order to deploy a new Private Link endpoint, we must take the following steps:

1. Go to the Azure portal and select the option to create a new service. Search for `Private Link` and select the **Create new** option.

2. In the new pane, **Private Link Center**, select **Create private endpoint**:

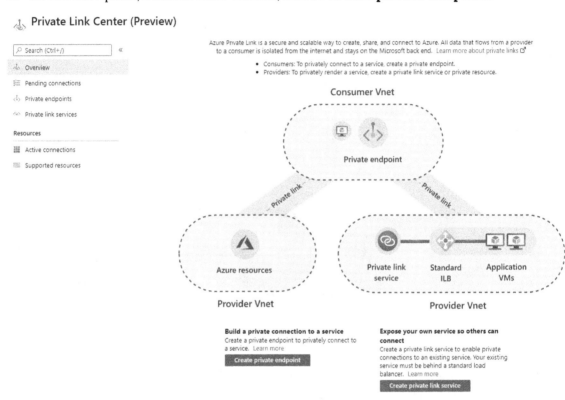

Figure 9.23: Creating a new Private Link endpoint

3.  In the new pane, under the **Basics** section, provide information for **Subscription**, **Resource group**, **Name**, and **Region**:

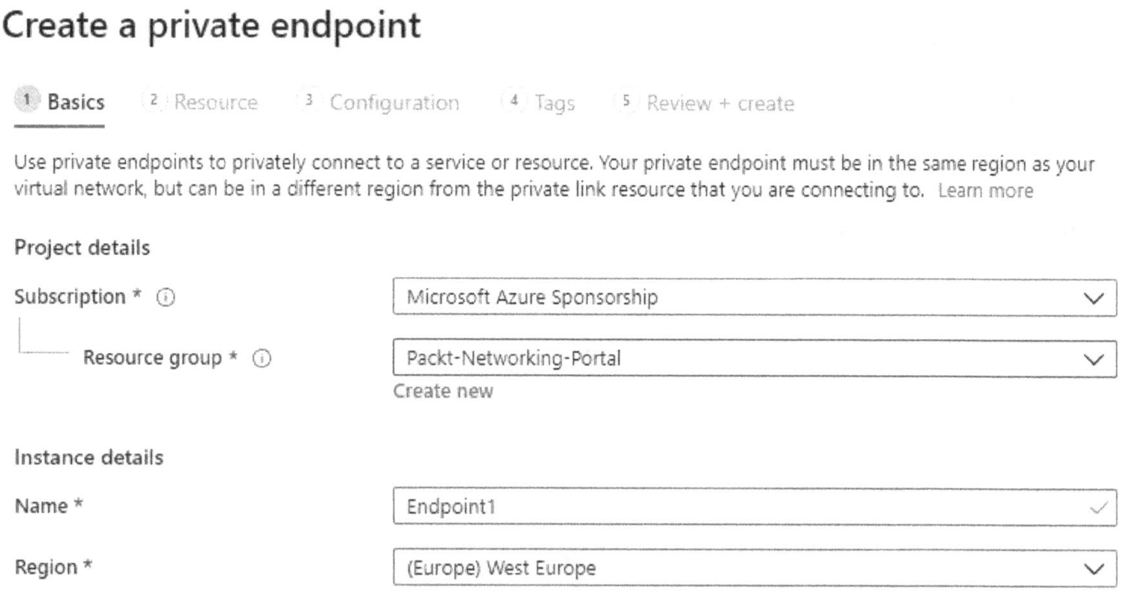

## Create a private endpoint

① **Basics**  ② Resource  ③ Configuration  ④ Tags  ⑤ Review + create

Use private endpoints to privately connect to a service or resource. Your private endpoint must be in the same region as your virtual network, but can be in a different region from the private link resource that you are connecting to.  Learn more

### Project details

Subscription * ⓘ

> Microsoft Azure Sponsorship                                          ⌄

Resource group * ⓘ

> Packt-Networking-Portal                                               ⌄

Create new

### Instance details

Name *

> Endpoint1                                                             ✓

Region *

> (Europe) West Europe                                                  ⌄

Figure 9.24: Basic information for the Private Link endpoint

4.  In the **Resource** section, we must select an option for **Subscription**, **Resource type** (in our case, **Microsoft.Sql/servers**), **Resource** (only resources of the selected resource type will be available), and **Target sub-resource**:

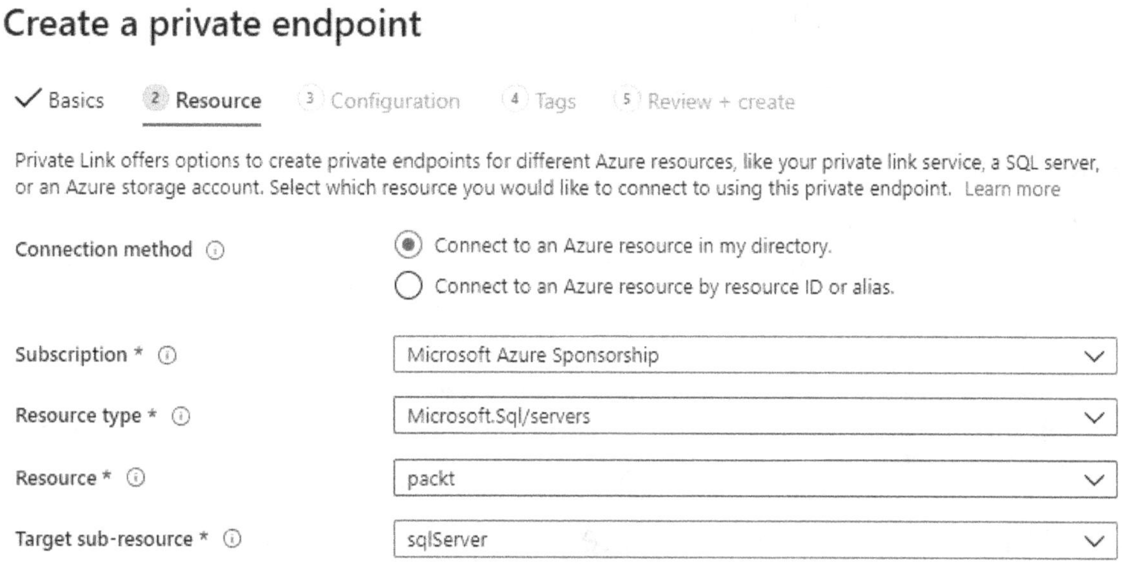

## Create a private endpoint

✓ Basics  ② **Resource**  ③ Configuration  ④ Tags  ⑤ Review + create

Private Link offers options to create private endpoints for different Azure resources, like your private link service, a SQL server, or an Azure storage account. Select which resource you would like to connect to using this private endpoint.  Learn more

Connection method ⓘ
>  ◉ Connect to an Azure resource in my directory.
>  ○ Connect to an Azure resource by resource ID or alias.

Subscription * ⓘ

> Microsoft Azure Sponsorship                                          ⌄

Resource type * ⓘ

> Microsoft.Sql/servers                                                 ⌄

Resource * ⓘ

> packt                                                                 ⌄

Target sub-resource * ⓘ

> sqlServer                                                             ⌄

Figure 9.25: Configuring the resources for the Private Link endpoint

5. On the **Configuration** pane, we must provide **Networking** settings and select the virtual network and subnet that will be associated. Optionally, we can add integration with a private DNS. If we choose to add DNS integration, we must provide information for **Subscription** and **Private DNS zones**:

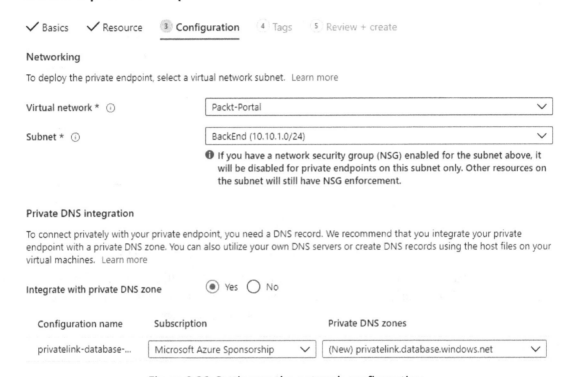

Figure 9.26: Setting up the network configuration

## How it works...

The Private Link endpoint associates the selected PaaS resource with the subnet on the virtual network. By doing so, we have the option to access the PaaS resource over a secure connection. Optionally, we can integrate a private DNS zone and use DNS resolution instead of IP addresses.

A Private Link endpoint allows us to link services directly but only individual services and only directly. If we need to add load balancers in place, we can use a Private Link service.

## Creating a Private Link service

A Private Link service allows us to set up a secure connection to resources associated with Standard Load Balancer. For that, we need to prepare infrastructure prior to deploying the Private Link service.

## Getting ready

We must create a virtual machine first. Check the *Creating Azure virtual machines* recipe from *Chapter 2, Virtual machine networking*. Note that in the **Networking** section, we want to select the same virtual network that was used to connect the SQL server in the previous recipe.

A Private Link service requires Standard Load Balancer as well. See the *Creating a public load balancer*, *Creating a backend pool*, *Creating health probes*, and *Creating load balancer rules* recipes from *Chapter 10, Load balancers*. Note that in the backend target, we need to select the virtual machine we just created.

Now, open the browser and go to the Azure portal via https://portal.azure.com.

## How to do it...

In order to deploy the new Private Link service, we must take the following steps:

1. In the Azure portal, select the option to create a new service. Search for `Private Link` and select the **Create new** option.

2. In the new pane, **Private Link Center**, select **Create private link service**:

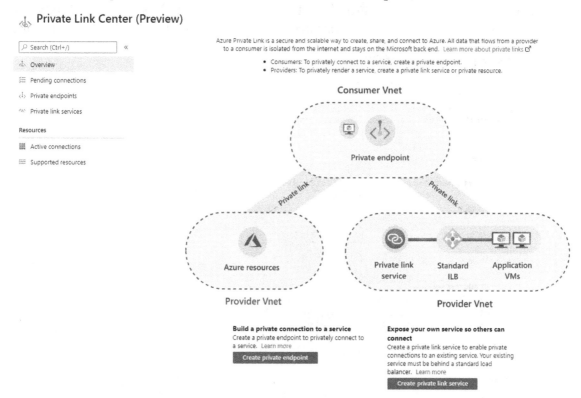

**Figure 9.27: Creating a new Private Link service**

3. Under **Basics**, we need to provide information for **Subscription**, **Resource group**, **Name**, and **Region**:

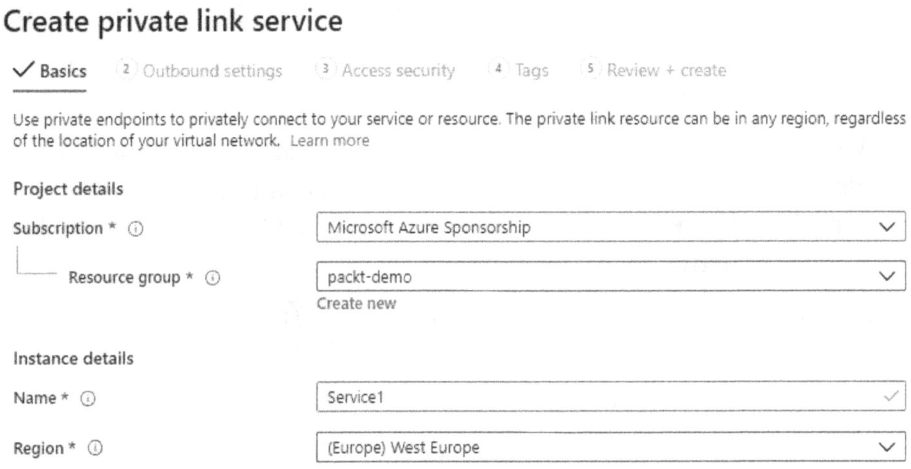

### Create private link service

✓ **Basics**    2  Outbound settings    3  Access security    4  Tags    5  Review + create

Use private endpoints to privately connect to your service or resource. The private link resource can be in any region, regardless of the location of your virtual network. Learn more

**Project details**

| | |
|---|---|
| Subscription * ⓘ | Microsoft Azure Sponsorship ⌄ |
| └─ Resource group * ⓘ | packt-demo ⌄ |
| | Create new |

**Instance details**

| | |
|---|---|
| Name * ⓘ | Service1 ✓ |
| Region * ⓘ | (Europe) West Europe ⌄ |

Figure 9.28: Information about the new Private Link service

4. Under **Outbound settings**, we must select options for **Load Balancer**, **Load Balancer frontend IP address**, and **Source NAT subnet**. **Source NAT Virtual network** is automatically selected and grayed out. We can also select **Yes** or **No** for **Enable TCP proxy V2** and whether the private IP address is going to be dynamic or static:

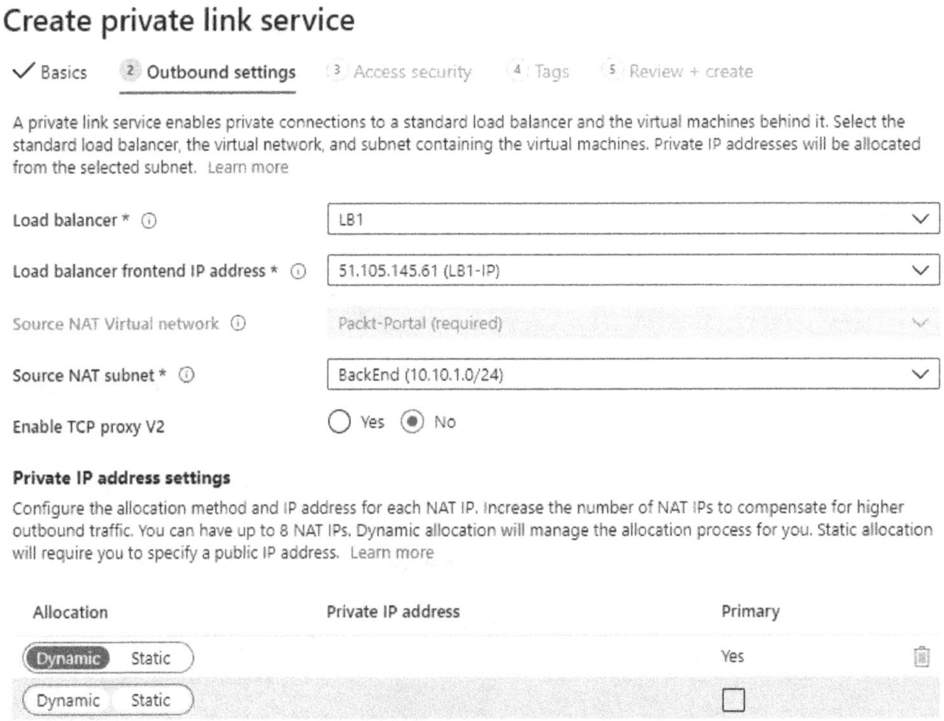

### Create private link service

✓ Basics    2  **Outbound settings**    3  Access security    4  Tags    5  Review + create

A private link service enables private connections to a standard load balancer and the virtual machines behind it. Select the standard load balancer, the virtual network, and subnet containing the virtual machines. Private IP addresses will be allocated from the selected subnet. Learn more

| | |
|---|---|
| Load balancer * ⓘ | LB1 ⌄ |
| Load balancer frontend IP address * ⓘ | 51.105.145.61 (LB1-IP) ⌄ |
| Source NAT Virtual network ⓘ | Packt-Portal (required) ⌄ |
| Source NAT subnet * ⓘ | BackEnd (10.10.1.0/24) ⌄ |
| Enable TCP proxy V2 | ◯ Yes  ◉ No |

**Private IP address settings**

Configure the allocation method and IP address for each NAT IP. Increase the number of NAT IPs to compensate for higher outbound traffic. You can have up to 8 NAT IPs. Dynamic allocation will manage the allocation process for you. Static allocation will require you to specify a public IP address. Learn more

| Allocation | Private IP address | Primary | |
|---|---|---|---|
| **Dynamic**  Static | | Yes | 🗑 |
| Dynamic  Static | | ☐ | |

Figure 9.29: Configuring the outbound settings

5. In **Access security**, we can select who can request access to our service. The options are **Role-based access control only (RBAC)**, **Restricted by subscription**, and **Anyone with your alias**. The default and recommended option is to use RBAC as native access control in Azure:

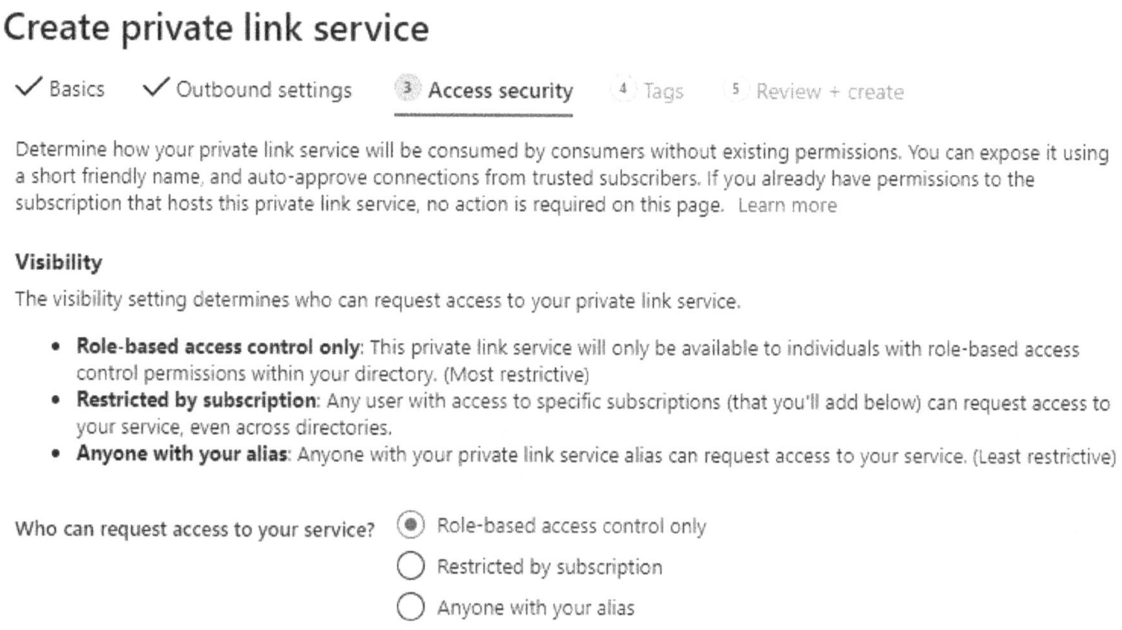

# Create private link service

✓ Basics   ✓ Outbound settings   ③ Access security   ④ Tags   ⑤ Review + create

Determine how your private link service will be consumed by consumers without existing permissions. You can expose it using a short friendly name, and auto-approve connections from trusted subscribers. If you already have permissions to the subscription that hosts this private link service, no action is required on this page.  Learn more

**Visibility**

The visibility setting determines who can request access to your private link service.

- **Role-based access control only**: This private link service will only be available to individuals with role-based access control permissions within your directory. (Most restrictive)
- **Restricted by subscription**: Any user with access to specific subscriptions (that you'll add below) can request access to your service, even across directories.
- **Anyone with your alias**: Anyone with your private link service alias can request access to your service. (Least restrictive)

Who can request access to your service?   ⦿ Role-based access control only

⦾ Restricted by subscription

⦾ Anyone with your alias

**Figure 9.30: The Access security pane**

## How it works...

A Private Link service and a Private Link endpoint work in a similar way, allowing us to connect to services (that are by default publicly accessible) over a private network. The main difference is that with a Private Link endpoint, we link PaaS services, and with a Private Link service, we create a custom service behind Standard Load Balancer.

# 10

# Load balancers

**Load balancers** are used to support scaling and high availability for applications and services. A load balancer is primarily composed of three components—a frontend, a backend, and routing rules. Requests coming to the frontend of a load balancer are distributed based on routing rules to the backend, where we place multiple instances of a service. This can be used for performance-related reasons, where we would like to distribute traffic equally between endpoints in the backend, or for high availability, where multiple instances of services are used to increase the chances that at least one endpoint will be available at all times.

We will cover the following recipes in this chapter:

- Creating an internal load balancer
- Creating a public load balancer
- Creating a backend pool
- Creating health probes
- Creating load balancer rules
- Creating inbound NAT rules
- Creating explicit outbound rules

# Technical requirements

For this chapter, an Azure subscription is required.

The code samples can be found at https://github.com/PacktPublishing/Azure-Networking-Cookbook-Second-Edition/tree/master/Chapter10.

# Creating an internal load balancer

Microsoft Azure supports two types of load balancers—**internal** and **public**. An internal load balancer is assigned a private IP address (from the address range of subnets in the virtual network) for a frontend IP address, and it targets the private IP addresses of our services (usually, an Azure **virtual machine** (**VM**)) in the backend. An internal load balancer is usually used by services that are not internet-facing and are accessed only from within our virtual network.

## Getting ready

Before you start, open the browser and go to the Azure portal via https://portal.azure.com.

## How to do it...

In order to create a new internal load balancer with the Azure portal, we must use the following steps:

1.  In the Azure portal, select **Create a resource** and choose **Load Balancer** under **Networking** services (or search for `Load Balancer` in the search bar).

2.  In the new pane, we must select a **Subscription** option and a **Resource group** option for where the load balancer is to be created. Then, we must provide information for the **Name**, **Region**, **Type**, and **SKU** options. In this case, we select **Internal** for **Type** to deploy an internal load balancer and set **SKU** to **Standard**. Finally, we must select the **Virtual network** and the **Subnet** that the load balancer will be associated with, along with information about the **IP address assignment**, which can be **Static** or **Dynamic**:

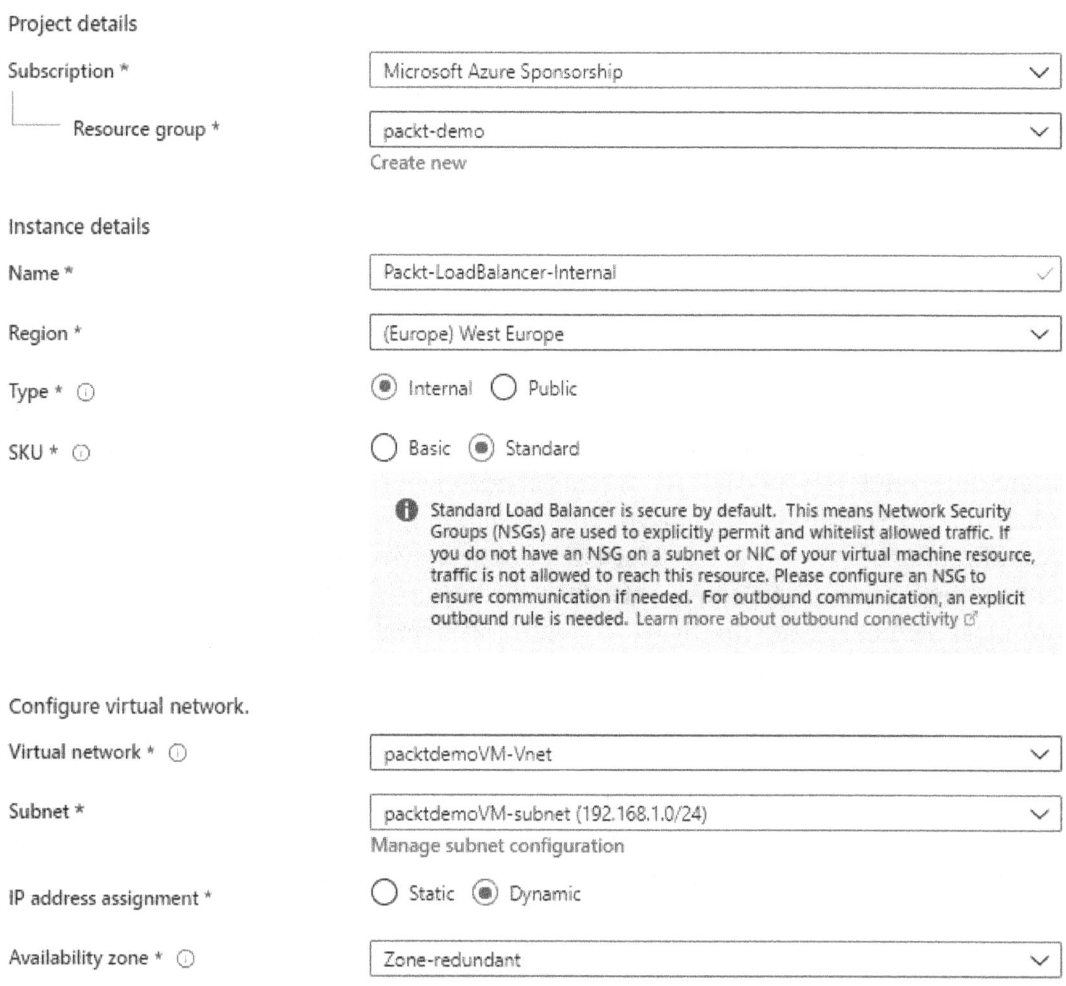

Project details

Subscription *

> Microsoft Azure Sponsorship ⌄

Resource group *

> packt-demo ⌄

Create new

Instance details

Name *

> Packt-LoadBalancer-Internal ✓

Region *

> (Europe) West Europe ⌄

Type * ⓘ

( ● ) Internal    ( ) Public

SKU * ⓘ

( ) Basic    ( ● ) Standard

> ⓘ   Standard Load Balancer is secure by default. This means Network Security Groups (NSGs) are used to explicitly permit and whitelist allowed traffic. If you do not have an NSG on a subnet or NIC of your virtual machine resource, traffic is not allowed to reach this resource. Please configure an NSG to ensure communication if needed. For outbound communication, an explicit outbound rule is needed. Learn more about outbound connectivity ⌐

Configure virtual network.

Virtual network * ⓘ

> packtdemoVM-Vnet ⌄

Subnet *

> packtdemoVM-subnet (192.168.1.0/24) ⌄

Manage subnet configuration

IP address assignment *

( ) Static    ( ● ) Dynamic

Availability zone * ⓘ

> Zone-redundant ⌄

**Figure 10.1: Creating a new internal load balancer**

3. After all the information is entered, we select the **Review + create** option to validate the information and start the deployment of the load balancer.

## How it works...

An internal load balancer is assigned a private IP address, and all requests coming to the frontend of an internal load balancer must come to that private address. This limits the traffic coming to the load balancer to be from within the virtual network associated with the load balancer. Traffic can come from other networks (other virtual networks or local networks) if there is some kind of **virtual private network (VPN)** in place. The traffic coming to the frontend of the internal load balancer will be distributed across the endpoints in the backend of the load balancer. Internal load balancers are usually used for services that are not placed in a **demilitarized zone (DMZ)** (and are therefore not accessible over the internet), but rather in middle- or back-tier services in a multi-tier application architecture.

We also need to keep in mind the differences between the **Basic** and **Standard** SKUs. The main difference is in performance (this is better in the Standard SKU) and SLA (Standard has an SLA guaranteeing 99.99% availability, while Basic has no SLA). Also, note that Standard SKU requires a **Network Security Group (NSG)**. If an NSG is not present on the subnet or **Network Interface**, or **NIC** (of the VM in the backend), traffic will not be allowed to reach its target. For more information on load balancer SKUs, see https://docs.microsoft.com/azure/load-balancer/skus.

# Creating a public load balancer

The second type of load balancer in Azure is a **public load balancer**. The main difference is that a public load balancer is assigned a public IP address in the frontend, and all requests come over the internet. The requests are then distributed to the endpoints in the backend.

## Getting ready

Before you start, open the browser and go to the Azure portal via https://portal.azure.com.

## How to do it...

In order to create a new public load balancer with the Azure portal, we must follow these steps:

1.  In the Azure portal, select **Create a resource** and choose **Load Balancer** under **Networking** services (or search for `Load Balancer` in the search bar).

2. In the new pane, we must select a **Subscription** option and a **Resource group** option for where the load balancer is to be created. Then, we must provide information for **Name**, **Region**, **Type**, and **SKU**. In this case, we select **Public** for **Type** to deploy a public load balancer. and set **SKU** to **Standard**. Selecting **Public** as the load balancer type will slightly change the pane. We will no longer have the option to select a virtual network and subnet, as we did for the internal load balancer. Instead, we can choose options for **Public IP address** (new or existing), **Public IP address SKU**, IP address assignment, and whether we want to use IPv6. Note that the public IP address SKU depends directly on the load balancer SKU, so the SKU selected for the load balancer will transfer automatically to the IP address:

## Create load balancer

| | |
|---|---|
| Subscription * | Microsoft Azure Sponsorship |
| Resource group * | packt-demo |
| | Create new |

**Instance details**

| | |
|---|---|
| Name * | Packt-LoadBalancer-Public |
| Region * | (Europe) West Europe |
| Type * ⓘ | ○ Internal  ⦿ Public |
| SKU * ⓘ | ○ Basic  ⦿ Standard |

ⓘ Standard Load Balancer is secure by default. This means Network Security Groups (NSGs) are used to explicitly permit and whitelist allowed traffic. If you do not have an NSG on a subnet or NIC of your virtual machine resource, traffic is not allowed to reach this resource. Please configure an NSG to ensure communication if needed. For outbound communication, an explicit outbound rule is needed. Learn more about outbound connectivity ⧉

**Public IP address**

| | |
|---|---|
| Public IP address * ⓘ | ⦿ Create new  ○ Use existing |
| Public IP address name * | Packt-LoadBalancer-PublicIP |
| Public IP address SKU | Standard |
| Assignment | ○ Dynamic  ⦿ Static |
| Availability zone * | Zone-redundant |
| Add a public IPv6 address ⓘ | [ No ] Yes |

Figure 10.2: Creating a new public load balancer

3.  After all the information is entered, select the **Review + create** option to validate the information and start the deployment of the load balancer.

## How it works...

The public load balancer is assigned a public IP address in the frontend. Therefore, all requests coming to the public load balancer will come over the internet, targeting the load balancer's public IP address. Requests are then distributed to endpoints in the backend of the load balancer. What's interesting is that the public load balancer does not target the public IP addresses in the backend, but private IP addresses instead. For example, let's say that we have one public load balancer with two Azure VMs in the backend. Traffic coming to the public IP address of the load balancer will then be distributed to VMs but will target the VMs' private IP addresses.

Public load balancers are used for public-facing services, most commonly for web servers.

# Creating a backend pool

After the load balancer is created, either internally or publicly, we need to configure it further in order to start using it. During the creation process, we define the frontend of the load balancer and know where traffic needs to go to reach the load balancer. But, in order to define where that traffic needs to go after reaching the load balancer, we must first define a backend pool.

## Getting ready

Before you start, open the browser and go to the Azure portal via https://portal.azure.com.

## How to do it...

In order to create the backend pool, we must do the following:

1.  In the Azure portal, locate the previously created load balancer (either internal or public).

2.  In the **Load balancer** pane, under **Settings**, select **Backend pools**. Select **Add** to add the new backend pool:

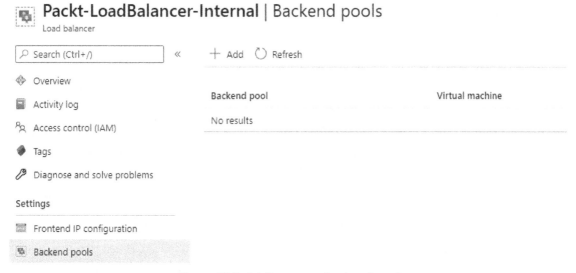

Figure 10.3: Adding a new backend pool

3. In the new pane, we must provide a **Name** and specify what the load balancer is associated to. Associations can be created for VMs or VM scale sets. In this example, we will use **Virtual machines**. Based on this selection, you will be offered additional options to add VMs to the backend pool:

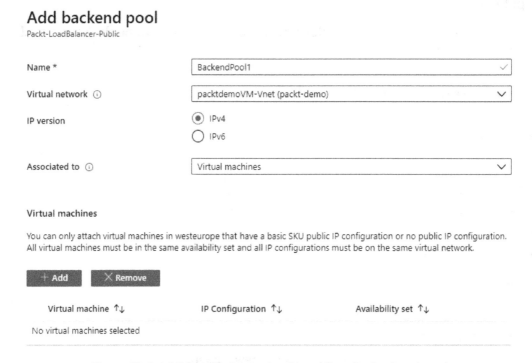

Figure 10.4: Additional information for adding the backend pool

4. Click **Add** and a new pane will open. Here we can add the VMs we want to associate with the backend pool. Note that the VMs must be in the same virtual network as the load balancer and in the same availability set. Select the VMs that you want to add to the backend pool:

## Add virtual machines to backend pool

Figure 10.5: Adding VMs to the backend pool

5. After the VMs are selected, they will appear under the **Virtual machines** list for creating the pool. Click **Add** to create the backend pool with the associated VMs:

Figure 10.6: List of VMs for creating a backend pool

6. After the configuration is entered, it takes a few minutes to create the backend pool. After that, the associated resources will show up in the backend pool list:

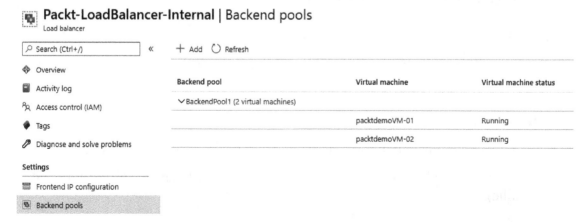

Figure 10.7: The backend pool list

## How it works...

The two main components of any load balancer are the frontend and the backend. The frontend defines the endpoint of the load balancer, and the backend defines where the traffic needs to go after reaching the load balancer. As the frontend information is created along with the load balancer, we must define the backend ourselves, after which the traffic will be evenly distributed across endpoints in the backend. The available options for the backend pool are VMs and VM scale sets.

## See also

More information on VMs, availability sets, and VM scale sets is available in my book, *Hands-On Cloud Administration in Azure*, published by Packt at https://www.packtpub.com/virtualization-and-cloud/hands-cloud-administration-azure.

# Creating health probes

After the frontend and the backend of the load balancer are defined, traffic is evenly distributed among endpoints in the backend. But what if one of the endpoints is unavailable? In that case, some of the requests will fail until we detect the issue, or even fail indefinitely should the issue remain undetected. The load balancer would send a request to all the defined endpoints in the backend pool and the request would fail when directed to an unavailable server.

This is why we introduce the next two components in the load balancer—**health probes** and **rules**. These components are used to detect issues and define what to do when issues are detected.

Health probes constantly monitor all endpoints defined in the backend pool and detect if any of them become unavailable. They do this by sending a probe in the configured protocol and listening for a response. If an HTTP probe is configured, an HTTP 200 OK response is required to be considered successful.

## Getting ready

Before you start, open the browser and go to the Azure portal via https://portal.azure.com.

## How to do it...

To create a new health probe in the load balancer, we must do the following:

1. In the Azure portal, locate the previously created load balancer (either internal or public).

2.  In the **Load balancer** pane, under **Settings**, select **Health probes**. Select **Add** to add a new health probe:

**Figure 10.8: Adding a new health probe**

3.  In the new pane, we need to provide information about the health probe's **Name** and IP version, or **Protocol**, we want to use, as well as configuring the **Port**, **Interval**, and **Unhealthy threshold** options, as shown in *Figure 10.9*:

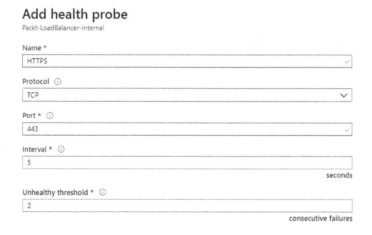

**Figure 10.9: Providing health probe information**

4.  After we select **OK**, the new health probe will be created and will appear on the list of available health probes associated with the load balancer.

## How it works...

After we define the health probe, it will be used to monitor the endpoints in the backend pool. We define the protocol and the port as useful information that will provide information regarding whether the service we are using is available or not. Monitoring the state of the server would not be enough, as it could be misleading. For example, the server could be running and available, but the IIS or SQL server that we use might be down. So, the protocol and the port will detect changes in the service that we are interested in and not only whether the server is running. The interval defines how often a check is performed, and the unhealthy threshold defines after how many consecutive fails the endpoint is declared unavailable.

# Creating load balancer rules

The last piece of the puzzle when speaking of Azure load balancers is the **rule**. Rules finally tie all things together and define which health probe (there can be more than one) will monitor which backend pool (more than one can be available). Furthermore, rules enable port mapping from the frontend of a load balancer to the backend pool, defining how ports relate and how incoming traffic is forwarded to the backend.

## Getting ready

Before you start, open your browser and go to the Azure portal via https://portal.azure.com.

## How to do it...

In order to create a load balancer rule, we must do the following:

1. In the Azure portal, locate the previously created load balancer (either internal or public).

2. In the **Load balancer** pane, under **Settings**, select **Load balancing rules**. Select **Add** to add a load balancing rule:

Figure 10.10: Adding load balancing rules

3. In the new pane, we must provide information for the **Name** and the **IP version** we are going to use, which **Frontend IP address** we are going to use (as a load balancer can have more than one), the **Protocol**, and the **Port** mapping (traffic from the incoming port will be forwarded to the backend port). If we enable high-availability ports (only available on internal load balancers), this will remove the protocol options and enable load balancing on all ports for TCP and UDP protocols. Furthermore, we need to provide information for the **Backend Port**, **Backend pool**, **Health probe**, **Session persistence**, and **Idle timeout** (**minutes**) settings, and decide whether we want to use a **Floating IP**. Finally, we have the option to create an implicit outbound rule:

### Add load balancing rule
Packt-LoadBalancer-Public

Name *

    Rule1

IP Version *
◉ IPv4  ○ IPv6

Frontend IP address * ○

    192.168.1.6 (LoadBalancerFrontEnd)

☐ HA Ports ○

Protocol
◉ TCP  ○ UDP

Port *

    443

Backend port * ○

    443

Backend pool ○

    BackendPool1 (2 virtual machines)

Health probe ○

    HTTPS (TCP:443)

Session persistence ○

    None

Idle timeout (minutes) ○

    4

TCP reset
◉ Disabled  ○ Enabled

Floating IP (direct server return) ○
( Disabled  Enabled )

Create implicit outbound rules ○
◉ Yes  ○ No

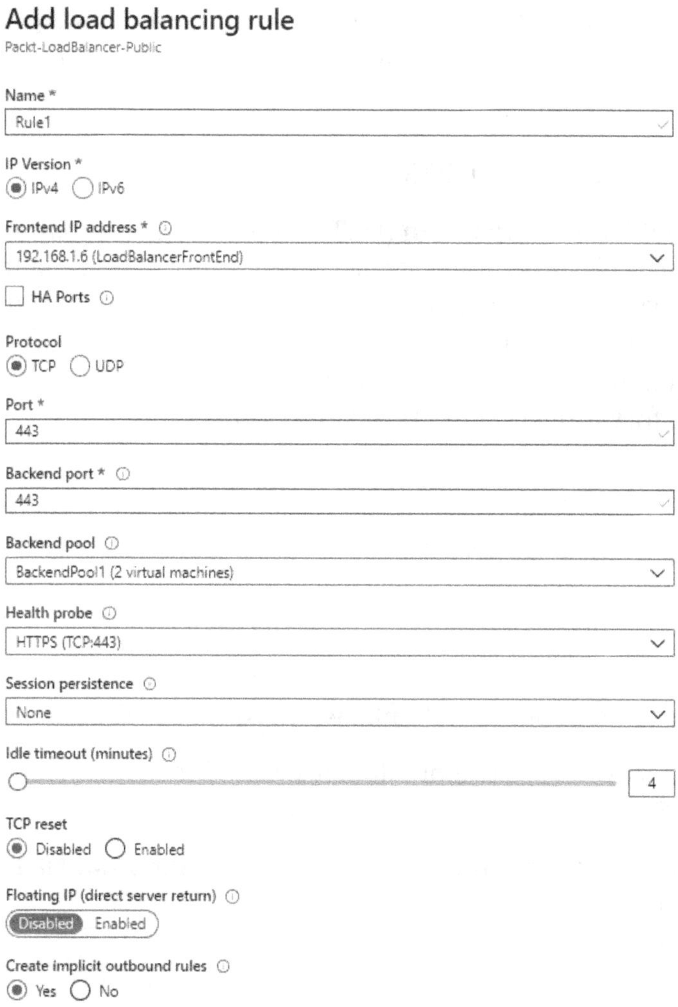

Figure 10.11: Configuring load balancing rules

4. After we select **OK**, a new rule will be created, which will appear on the list of available load balancing rules.

## How it works...

The load balancer rule is the final piece that ties all the components together. We define which frontend IP address is used and which backend the pool traffic will be forwarded to. The health probe is assigned to monitor the endpoints in the backend pool and to keep track of whether there are any unresponsive endpoints. We also create a port mapping that will determine which protocol and port the load balancer will listen on and, when the traffic arrives, where this traffic will be forwarded.

As its default distribution mode, Azure Load Balancer uses a five-tuple hash (source IP, source port, destination IP, destination port, and protocol type). If we change the session persistence to **Client IP**, the distribution will be two-tuple (requests from the same client IP address will be handled by the same VM). Changing session persistence to **Client IP and protocol** will change the distribution to three-tuple (requests from the same client IP address and protocol combination will be handled by the same VM).

# Creating inbound NAT rules

Inbound **Network Address Translation** (**NAT**) rules are an optional setting in Azure Load Balancer. These rules essentially create another port mapping from the frontend to the backend, forwarding traffic from a specific port on the frontend to a specific port in the backend. The difference between inbound NAT rules and port mapping in load balancer rules is that inbound NAT rules apply to direct forwarding to a VM, whereas load balancer rules forward traffic to a backend pool.

## Getting ready

Before you start, open the browser and go to the Azure portal via https://portal.azure. com.

## How to do it...

In order to create a new inbound NAT rule, we must do the following:

1. In the Azure portal, locate the previously created load balancer (either internal or public).

2. In the **Load balancer** pane, under **Settings**, select **Inbound NAT rules**. Select **Add** to add a new inbound NAT rule:

Figure 10.12: Adding an inbound NAT rule for an existing load balancer

3. In the new pane, we must provide details for the **Name**, **Frontend IP address**, **IP Version** (set based on the frontend IP address), **Service**, **Protocol**, and **Port** fields. We can also edit **Idle timeout**, which is set to **4** minutes by default. Select **Target virtual machine** and **Network IP configuration** for the same machine (if the VM has more than one IP configuration). Finally, you can select the default port mapping or use a custom one:

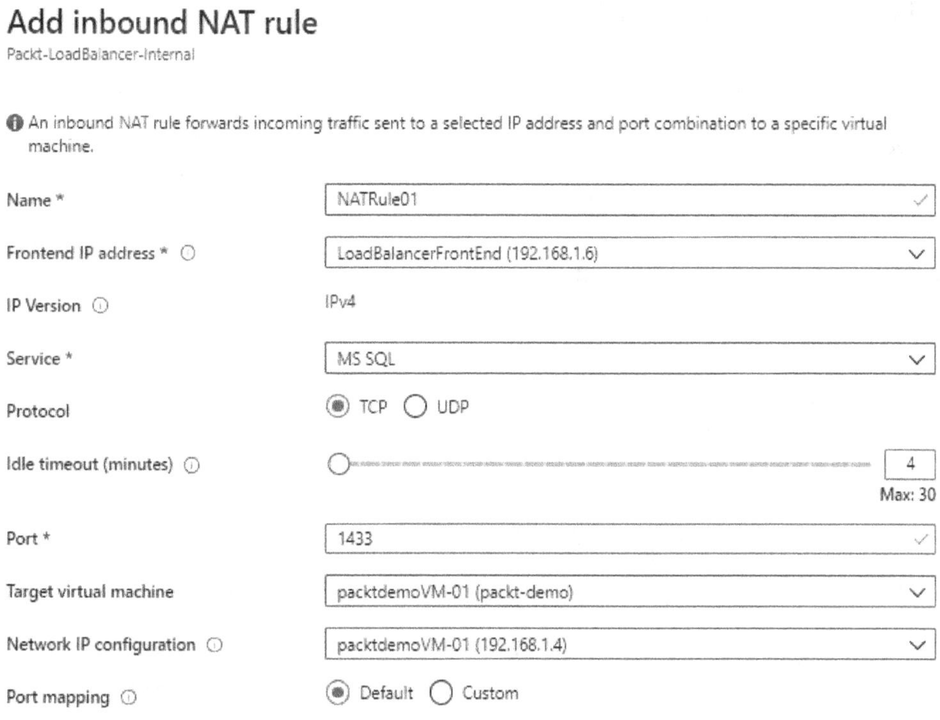

## Add inbound NAT rule

Packt-LoadBalancer-Internal

🛈 An inbound NAT rule forwards incoming traffic sent to a selected IP address and port combination to a specific virtual machine.

| | |
|---|---|
| Name * | NATRule01 |
| Frontend IP address * ⓘ | LoadBalancerFrontEnd (192.168.1.6) |
| IP Version ⓘ | IPv4 |
| Service * | MS SQL |
| Protocol | ⦿ TCP  ◯ UDP |
| Idle timeout (minutes) ⓘ | 4  Max: 30 |
| Port * | 1433 |
| Target virtual machine | packtdemoVM-01 (packt-demo) |
| Network IP configuration ⓘ | packtdemoVM-01 (192.168.1.4) |
| Port mapping ⓘ | ⦿ Default  ◯ Custom |

Figure 10.13: Configuring the inbound NAT rule settings

4. After we select **OK**, a new inbound NAT rule will be created.

## How it works...

Inbound NAT rules enable you to use the public IP of the load balancer to connect directly to a specific backend instance. They create a port mapping similar to the port mapping created by load balancer rules but to a specific backend instance. A load balancer rule creates additional settings, such as the health probe or session persistence. Inbound NAT rules exclude these settings and create unconditional mapping from the frontend to the backend. With an inbound NAT rule, forwarded traffic will always reach the single server in the backend, whereas a load balancer will forward traffic to the backend pool and will use a pseudo-round-robin algorithm to route traffic to any of the healthy servers in the backend pool.

# Creating explicit outbound rules

When creating load balancing rules, we can create implicit outbound rules. This will enable **Source Network Address Translation (SNAT)** for VMs in the backend pool and allow them to access the internet over the load balancer's public IP address (specified in the rule). But in some scenarios, implicit rules are not enough and we need to create explicit outbound rules. Explicit outbound rules (and SNAT in general) are available only for public load balancers with the Standard SKU.

## Getting ready

Before we begin, make sure that implicit outbound rules are disabled from load balancing rules:

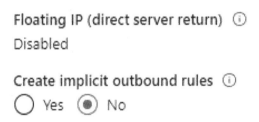

**Figure 10.14: Disabling implicit outbound rules**

Now, open the browser and go to the Azure portal via https://portal.azure.com.

## How to do it...

In order to create a load balancer rule, we must do the following:

1. In the Azure portal, locate the previously created public load balancer.

2. In the **Load balancer** pane, under **Settings**, select **Outbound rules**. Select **Add** to add the load balancing rule:

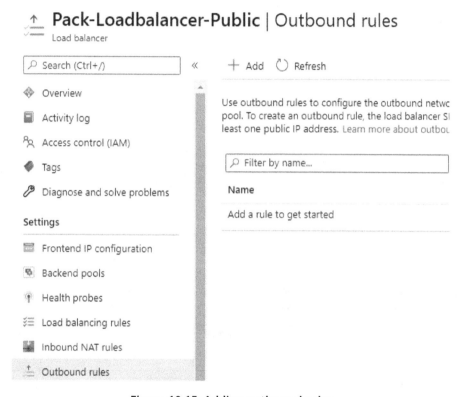

**Figure 10.15: Adding outbound rules**

3. In the **Outbound rules** pane, we must provide the rule name and select options for the **Frontend IP address**, **Protocol** (**All**, **TCP**, or **UDP**), **Idle timeout**, **TCP reset**, and **Backend pool** fields. In the **Port allocation** section of the same pane, we must select options for **Port allocation**, **Outbound ports**, **Ports per instance** (disabled when the maximum number of backend instances is selected), and **Maximum number of backend instances**:

## Add outbound rule
Pack-Loadbalancer-Public

| | |
|---|---|
| Name * | OutRule1 |
| Frontend IP address * ⓘ | 1 selected |
| | Create new |
| Protocol | ◉ All  ○ TCP  ○ UDP |
| Idle timeout (minutes) ⓘ | 4  Max: 30 |
| TCP Reset ⓘ | ◉ Enabled  ○ Disabled |
| Backend pool * ⓘ | BackendPool1 (2 instances) |
| | Create new |

**Port allocation**

Azure automatically assigns the number of outbound ports to use for source network address translation (SNAT) based on the number of frontend IP addresses and backend pool instances. Learn more about outbound connectivity ⧉

| | |
|---|---|
| Port allocation ⓘ | Manually choose number of outbound ports |
| Outbound ports | |
| Choose by * | Maximum number of backend instances |
| Ports per instance ⓘ | 0 |
| Frontend IPs | 1 |
| Maximum number of backend instances ⓘ | 2 |

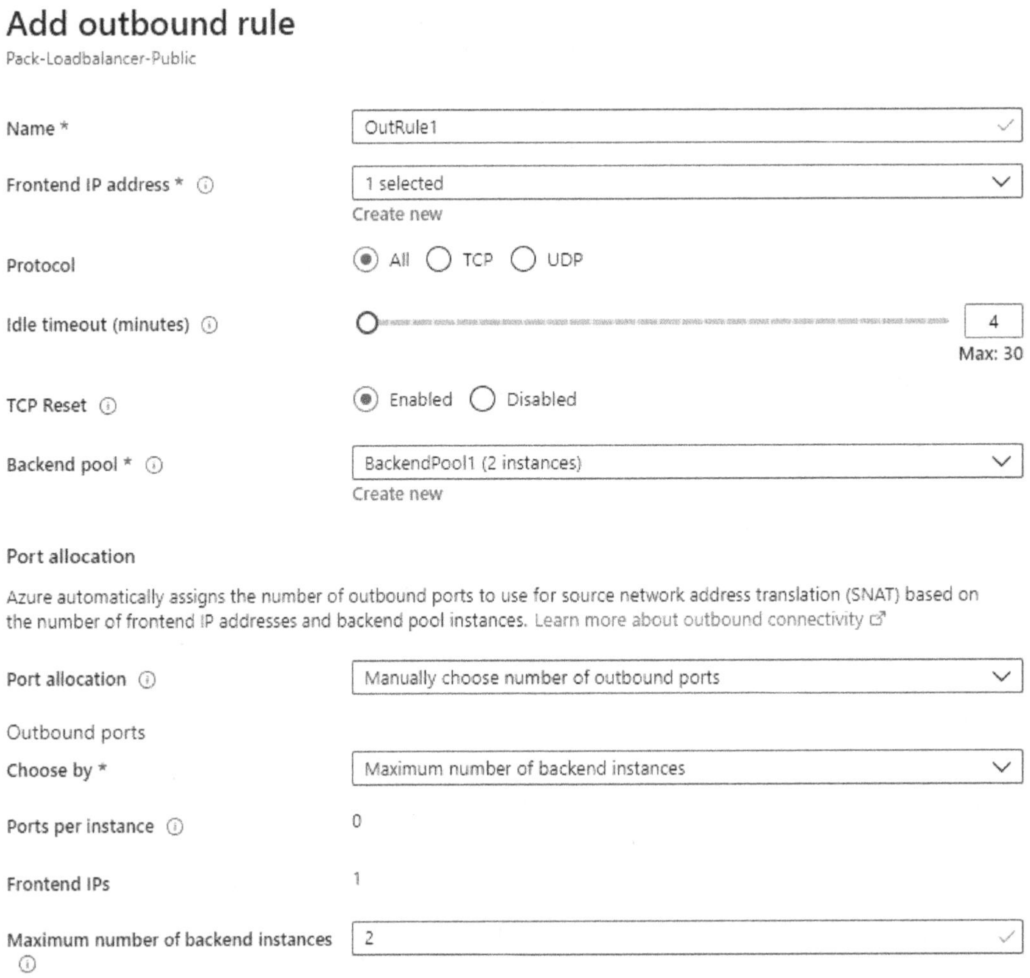

Figure 10.16: The outbound rule pane

## How it works...

Outbound rules depend on three things—frontend IP addresses, instances in the backend pool, and connections. Each frontend IP address has a limited number of ports for connections. The more IP addresses are assigned to the frontend, the more connections are allowed. On the other hand, the number of connections allowed (per backend instance) decreases with the number of instances in the backend.

If we set the default number of outbound ports, allocation is done automatically and without control. If we have a VM scale set with the default number of instances, port allocation will be done automatically for each VM in the scale set. If the number of instances in a scale set increases, this means that the number of ports allocated to each VM will drop in turn.

To avoid this, we can set port allocation to manual and either limit the number of instances that are allowed or limit the number of ports per instance. This will ensure that each VM has a certain number of ports dedicated and that connections will not be dropped.

# 11
# Traffic Manager

Azure Load Balancer is limited to providing high availability and scalability only to **Azure virtual machines (VMs)**. Also, a single load balancer is limited to VMs in a single Azure region. If we want to provide high availability and scalability to other Azure services that are globally distributed, we must introduce a new component–**Azure Traffic Manager**. Azure Traffic Manager is DNS-based and provides the ability to distribute traffic over services and spread traffic across Azure regions. But Traffic Manager is not limited to Azure services only; we can add external endpoints as well.

We will cover the following recipes in this chapter:

- Creating a new Traffic Manager profile
- Adding an endpoint
- Configuring distributed traffic
- Configuring traffic based on priority
- Configuring traffic based on geographical location
- Managing endpoints
- Managing profiles
- Configuring Traffic Manager with load balancers

# Technical requirements

For this chapter, an Azure subscription is required.

The code samples can be found at https://github.com/PacktPublishing/Azure-Networking-Cookbook-Second-Edition/tree/master/Chapter11.

# Creating a new Traffic Manager profile

Traffic Manager provides load balancing to services, but traffic is routed and directed using DNS entries. The front end is a **Fully Qualified Domain Name** (**FQDN**) assigned during creation, and all traffic coming to Traffic Manager is distributed to endpoints in the backend. In this recipe, we'll create a new Traffic Manager profile.

## Getting ready

Before you start, open your browser and go to the Azure portal via https://portal.azure.com.

## How to do it...

In order to create a new Traffic Manager profile, we must do the following:

1.  In the Azure portal, select **Create a resource** and choose **Traffic Manager Profile** under the **Networking** services (or search for `Traffic Manager Profile` in the search bar).

2.  In the new pane, we must provide information for the **Name**, **Routing method**, **Subscription**, and **Resource group** fields:

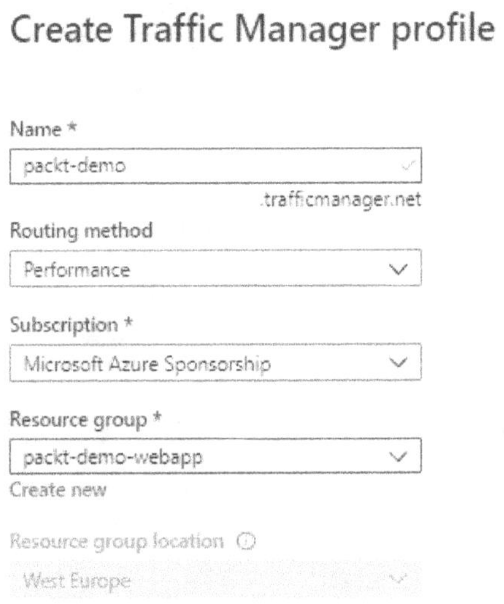

## Create Traffic Manager profile

Name *

packt-demo

.trafficmanager.net

Routing method

Performance

Subscription *

Microsoft Azure Sponsorship

Resource group *

packt-demo-webapp

Create new

Resource group location

West Europe

Figure 11.1: Providing information for a new Traffic Manager profile

3. Note that under the routing methods, we have multiple options to choose from– **Performance**, **Weighted**, **Priority**, **Geographic**, **MultiValue**, and **Subnet**. For this recipe, let's leave it as the default option (**Performance**), but we will cover the rest of the routing methods in other recipes in this chapter:

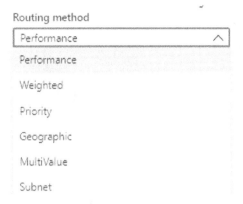

Figure 11.2: Selecting the Routing method

## How it works...

Traffic Manager is assigned a public endpoint that must be an FQDN. All traffic arriving at that endpoint will be distributed to endpoints in the backend, using the routing method selected. The default routing method is **Performance**. The performance method will distribute traffic based on the best possible performance available. For example, if we have more than one backend endpoint in the same region, traffic will be spread evenly. If the endpoints are located across different regions, Traffic Manager will direct traffic to the endpoint closest to the incoming traffic in terms of geographical location and minimum network latency.

Let's move on to the next recipe and add an endpoint to Traffic Manager.

# Adding an endpoint

After a Traffic Manager profile is created, we have the frontend endpoint and routing method defined. But we still need to define where the traffic needs to go after it's reached Traffic Manager. We need to add endpoints to the backend and define where the traffic is directed. In this recipe, we'll add a new endpoint to Traffic Manager.

## Getting ready

Before we can add endpoints to Traffic Manager, we need to create them. Executing the following script in PowerShell can help you create two web apps quickly:

```
$ResourceGroupName = "packt-demo-webapp"

$webappname="packt-demo-webapp"

$location1="West Europe"

$NumberOfWebApps= 2


New-AzResourceGroup -Name $ResourceGroupName '

-Location $location

$i=1

Do

{

New-AzWebApp -Name $webappname'-0'$i '

-Location $location '

-AppServicePlan $webappname '

-ResourceGroupName $ResourceGroupName

} While (($i=$I+1) -le $NumberOfWebApps)
```

The script can be edited to deploy more than two web apps if needed. However, to get the most out of Traffic Manager, it's best to have web apps in different regions.

After the script is completed, open your browser and go to the Azure portal at https://portal.azure.com.

## How to do it...

In order to add endpoints to Traffic Manager, we must do the following:

1.  In the Azure portal, locate the previously created Traffic Manager profile.

2.  In the **Traffic Manager profile** pane, under **Settings**, select **Endpoints**. Select **Add** to add a new endpoint:

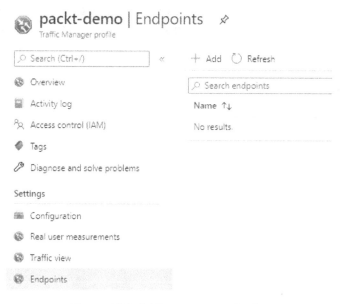

**Figure 11.3: Adding a new endpoint**

3. In the new pane, we need to provide information for the **Type** (of endpoint we are adding) and **Name** fields. For **Type**, we can choose between **Azure**, **External**, and **Nested**. If **Azure** is selected, we can select certain target resource types (**Cloud service**, **App service** or **slot**, and **Public IP address**), and based on the target resource type selection, we can select resources that fit the target resource type selected. Here, we have selected **packt-demo-webapp01**, which we created earlier:

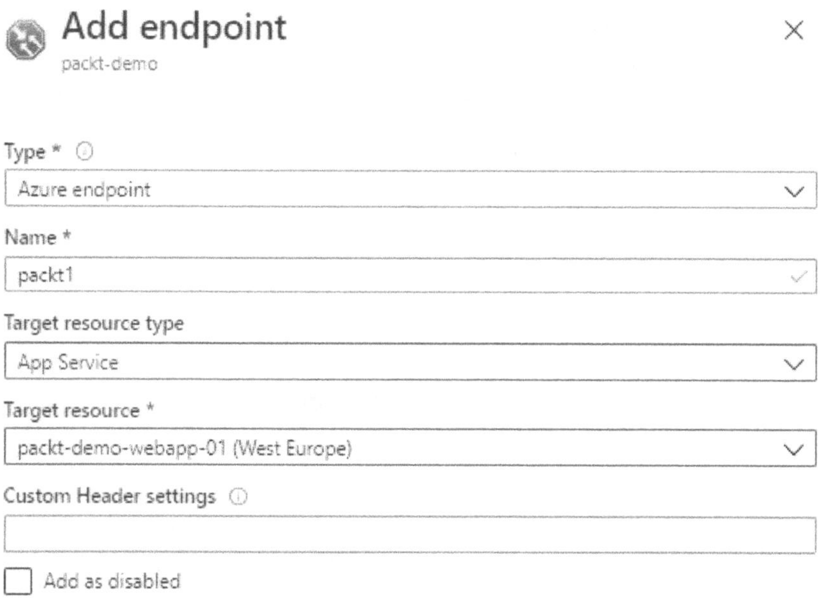

**Add endpoint**                                              ✕

packt-demo

Type *  ⓘ

| Azure endpoint                                    ⌄ |

Name *

| packt1                                            ✓ |

Target resource type

| App Service                                       ⌄ |

Target resource *

| packt-demo-webapp-01 (West Europe)                ⌄ |

Custom Header settings  ⓘ

| |

☐ Add as disabled

**Figure 11.4: Configuring the type of endpoint**

4. Adding a single endpoint will only work as a redirection from one FQDN to another. We need to repeat the process at least one more time and add at least one more endpoint:

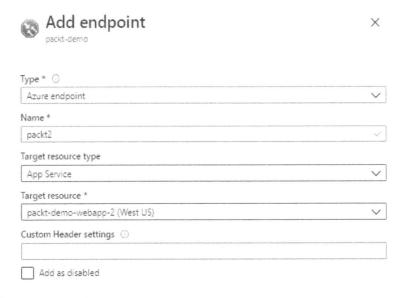

Figure 11.5: Adding a secondary endpoint

5. All the added endpoints will appear in the list of endpoints in the **Endpoint** section under the **Settings** option of Traffic Manager:

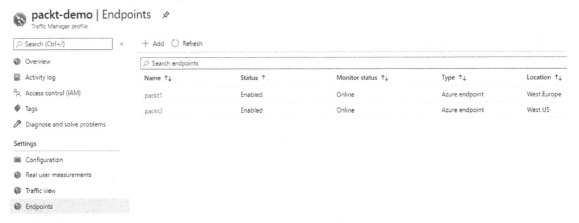

Figure 11.6: A list of endpoints

## How it works...

Incoming requests reach Traffic Manager by hitting the frontend endpoint of Traffic Manager. Based on rules (mainly the routing method), traffic is then forwarded to the backend endpoints. The load balancer works by forwarding traffic to private IP addresses. On the other hand, Traffic Manager uses public endpoints in the backend. The supported endpoint types are Azure, external, and nested. Based on the endpoint type, we can add Azure or external endpoints. Endpoints can be either (public) FQDNs or public IP addresses. Nested endpoints allow us to add other Traffic Manager profiles to the backend of the Traffic Manager.

Custom header settings add specific HTTP headers to the health checks that Traffic Manager sends to endpoints under a profile. They can be defined either at profile level (and applied to all endpoints under that profile) or for each individual endpoint. It comes in **header:value** format and we can add up to 8 pairs (**header1:value1**, **header2:value2**, **header3:value3**...)

After adding endpoints to Traffic Manager, let's move on to the next recipe and learn how to configure distributed traffic.

# Configuring distributed traffic

The default routing method for Traffic Manager is performance. The performance method will distribute traffic based on the best possible performance available. This method only takes full effect if we have multiple instances of a service in multiple regions. As this often isn't the case, other methods are available, such as distributed traffic (also referred to as the weighted routing method). In this recipe, we'll configure Traffic Manager to work in distributed mode.

## Getting ready

Before you start, open the browser and go to the Azure portal via https://portal.azure.com.

## How to do it...

In order to set distributed traffic, we must do the following:

1.  In the Azure portal, locate the previously created Traffic Manager profile.

2.  Under **Settings**, select the **Configuration** option. Here, we have multiple options that we can change, such as **DNS time to live (TTL)**, protocols, and failover settings:

**Figure 11.7: The Configuration pane for Traffic Manager**

3.  Change **Routing method** to **Weighted**, as shown in *Figure 11.8*. Furthermore, we can set up weight settings if needed:

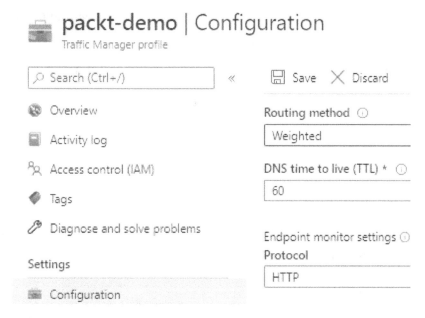

Figure 11.8: Changing the routing method to Weighted

## How it works...

The weighted routing method will distribute traffic evenly across all endpoints in the backend. We can further set weight settings to give an advantage to a certain endpoint and say that some endpoints will receive a bigger or smaller percentage of the traffic. This method is usually used when we have multiple instances of an application in the same region, or for scaling out to increase performance.

In this recipe, we learned how to distribute traffic evenly across all endpoints. In the next recipe, we'll learn how to configure traffic based on priority.

# Configuring traffic based on priority

Another routing method available is priority. Priority, as its name suggests, gives priority to some endpoints, while some endpoints are kept as backups. Backup endpoints are only used if endpoints with priority become unavailable. In this recipe, we'll configure Traffic Manager to route traffic based on priority.

## Getting ready

Before you start, open your browser and go to the Azure portal via https://portal.azure. com.

## How to do it...

In order to set the routing method to **Priority**, we must do the following:

1. In the Azure portal, locate the previously created **Traffic Manager profile**.

2. Under **Settings**, select the **Configuration** option.

3. Change **Routing method** to **Priority**, as shown in *Figure 11.9*:

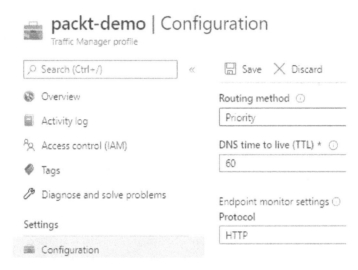

**Figure 11.9: Changing the routing method to Priority**

## How it works...

**Priority** sets a priority order for endpoints. All traffic will first go to the endpoints with the highest priority. Other endpoints (with lower priority) are backed up, and traffic is routed to these endpoints only when higher-priority endpoints become unavailable. The default priority order is the order of adding endpoints to Traffic Manager, where the endpoint added first becomes the one with the highest priority and the endpoint added last becomes the endpoint with the least priority. Priority can be changed under the endpoint settings.

In the next recipe, we will learn how to configure traffic based on geographical location.

# Configuring traffic based on geographical location

Geographical location is another routing method in Traffic Manager. This method is based on network latency and directs a request based on the geographical location of the origin and the endpoint. When a request comes to Traffic Manager, based on the origin of the request, it's routed to the nearest endpoint in terms of region. This way, it provides the least network latency possible. In this recipe, we'll configure Traffic Manager to route traffic based on geographical location.

## Getting ready

Before you start, open the browser and go to the Azure portal via https://portal.azure.com.

## How to do it...

In order to set the routing method to be based on geographical location, we must do the following:

1. In the Azure portal, locate the previously created Traffic Manager profile.

2. Under **Settings**, select the **Configuration** option.

3. Change the routing method to **Geographic**, as shown in *Figure 11.10*:

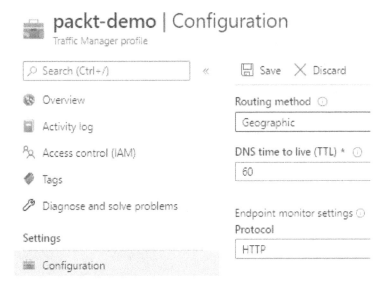

**Figure 11.10: Changing the routing method to Geographic**

## How it works...

The geographic routing method matches the request origin with the closest endpoint in terms of geographical location.

For example, let's say we have multiple endpoints, each on a different continent. If a request comes from Europe, it would make no sense to route it to Asia or North America. The geographic routing method will make sure that a request coming from Europe will be pointed to the endpoint located in Europe.

Let's move on to the next recipe and learn how to manage endpoints.

# Managing endpoints

After we add endpoints to Traffic Manager, we may have to make changes over time. This can be either to make adjustments or to completely remove endpoints. In this recipe, we'll edit existing Traffic Manager endpoints.

## Getting ready

Before you start, open the browser and go to the Azure portal via https://portal.azure.com.

## How to do it...

In order to make changes to endpoints in Traffic Manager, we must do the following:

1. In the Azure portal, locate the previously created Traffic Manager.

2. Under **Settings**, select **Endpoints**. From the list that appears, select the endpoint you want to change:

Figure 11.11: Changing endpoints in Traffic Manager

3. In the new pane, we can either delete, disable, or make adjustments to the endpoint:

### packt1
packt-demo

🖫 Save    ✕ Discard    🗑 Delete

Status
( Disabled  **Enabled** )

Monitor status
Online

Type
Azure endpoint

Target resource type

| App Service | ⌄ |

*Target resource

packt-demo-webapp-01    >

Custom Header settings ⓘ

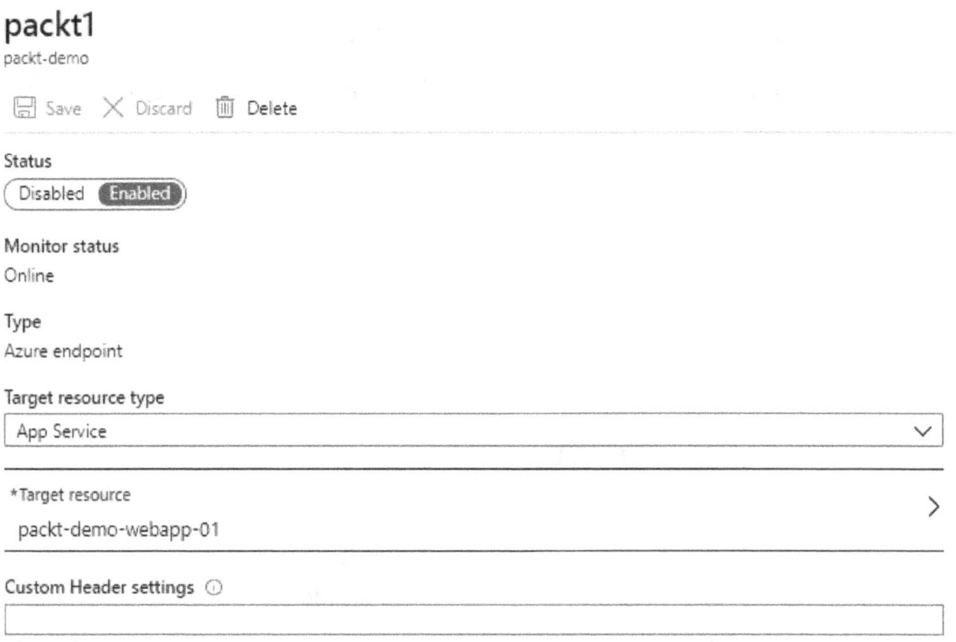

Figure 11.12: Pane for making adjustments to the endpoint

## How it works...

The existing endpoint in the Traffic Manager backend can be changed. We can delete the endpoint to completely remove it from Traffic Manager, or we can disable it to temporarily remove it from the backend. We can also change the endpoint completely, to point to another service or a completely different type.

In this recipe, we learned how to manage endpoints. In the next recipe, we will learn how to manage and adjust profiles.

# Managing profiles

The Traffic Manager profile is another setting that we can manage and adjust. Although it has very limited options, where we can only disable and enable Traffic Manager, managing the profile setting can be very useful for maintenance purposes. In this recipe, we'll manage our Traffic Manager profile.

## Getting ready

Before you start, open your browser and go to the Azure portal via https://portal.azure.com.

## How to do it...

In order to make changes to the Traffic Manager profile, we must do the following:

1.  In the Azure portal, locate the previously created Traffic Manager profile.

2.  In **Overview**, select the **Disable profile** option and confirm by clicking on the **Yes** button:

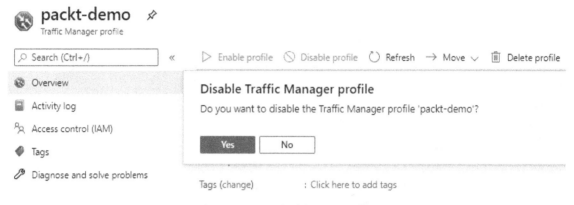

Figure 11.13: Disabling a profile

3.  Once the profile has been disabled, it can be enabled again with the **Enable profile** option:

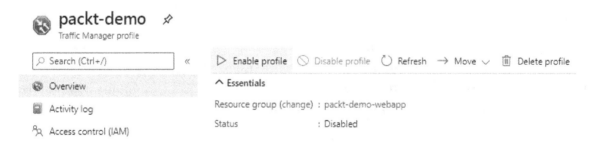

Figure 11.14: Enabling a profile

## How it works...

Managing the Traffic Manager profile with the disable and enable options will make the Traffic Manager frontend unavailable or available (based on the option selected). This can be very useful for maintenance purposes. If we must apply changes across all endpoints, and changes need to be applied to all endpoints at the same time, we can disable the Traffic Manager profile temporarily. Once the changes are applied to all the endpoints, we can make Traffic Manager available again by enabling the profile.

Let's move on to the next recipe and learn how to configure Traffic Manager with load balancers.

# Configuring Traffic Manager with load balancers

Combining Traffic Manager with load balancers is often done to provide maximum availability. Load balancers are limited to providing high availability to a set of resources located in the same region. This gives us an advantage if a single resource fails, as we have multiple instances of a resource. But *what if a complete region fails?* Load balancers can't handle resources in multiple regions, but we can combine load balancers with Traffic Manager to provide even better availability with resources across Azure regions. In this recipe, we'll configure Traffic Manager to work with load balancers.

## Getting ready

Before you start, open your browser and go to the Azure portal via https://portal.azure.com.

## How to do it...

In order to set up Traffic Manager with a load balancer, we must do the following:

1. In the Azure portal, locate the load balancer and verify that it has the assigned IP address as covered in *Chapter 8, Load balancers*. Only public IP addresses can be used:

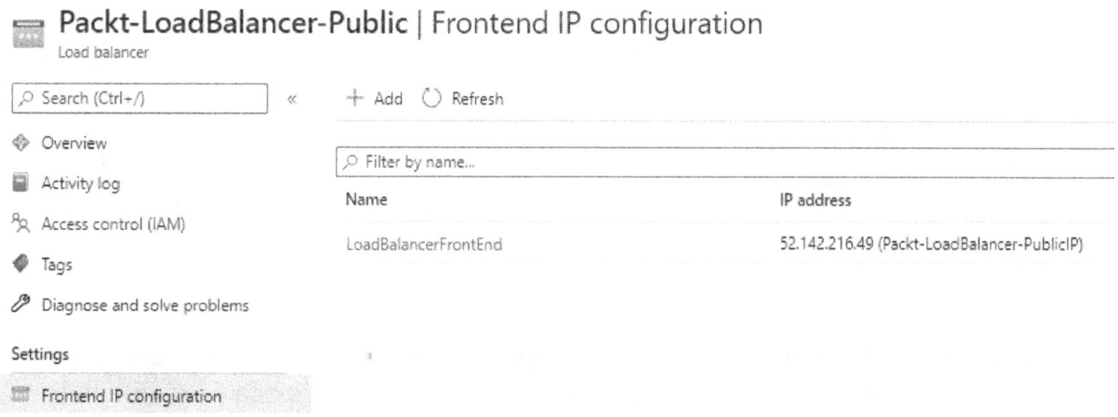

**Packt-LoadBalancer-Public** | Frontend IP configuration
Load balancer

| | | |
|---|---|---|
| Search (Ctrl+/) « | + Add ○ Refresh | |
| ◈ Overview | Filter by name... | |
| ▤ Activity log | | |
| | Name | IP address |
| ᏁᏁ Access control (IAM) | | |
| | LoadBalancerFrontEnd | 52.142.216.49 (Packt-LoadBalancer-PublicIP) |
| ◉ Tags | | |
| ℗ Diagnose and solve problems | | |
| Settings | | |
| ▦ Frontend IP configuration | | |

Figure 11.15: Verifying the assigned IP address of a load balancer

2. Go to Traffic Manager and select **Add** to add a new endpoint. Select **Azure endpoint** for **Type**, provide a name for the endpoint, and select **Public IP address** as the target resource type. Based on the type selected, a new option will appear, allowing us to select resources that match the type we selected. In our case, the option to select **Public IP address** is available:

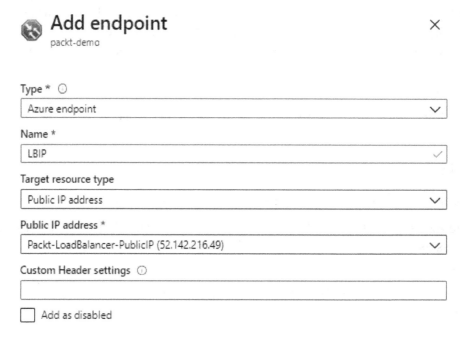

Figure 11.16: Configuring a new endpoint in Traffic Manager

3. Repeat the process and add another load balancer (from another region) as a second Traffic Manager endpoint.

## How it works...

Load balancers provide better availability, keeping a service active even if one of the services in the backend pool fails. If a region fails, load balancers can't provide help because they are limited to a single region. We must provide another set of resources in another region to truly increase availability—but these sets will be completely independent and will not provide failover unless we include Traffic Manager. Traffic Manager will become the frontend, and we will add load balancers as the backend endpoints of Traffic Manager. All requests will come to Traffic Manager first, and will then be routed to the appropriate load balancer in the backend. Traffic Manager will monitor the health of the load balancers, and if one of them becomes unavailable, the traffic will be rerouted to an active load balancer.

# 12

# Azure Application Gateway and Azure WAF

**Azure Application Gateway** is essentially a load balancer for web traffic, but it also provides us with better traffic control. Traditional load balancers operate on the transport layer and allow us to route traffic based on protocol (TCP or UDP) and IP address, mapping IP addresses, and protocols in the frontend to IP addresses and protocols in the back end. This "classic" operation mode is often referred to as layer 4. Application gateway expands on that and allows us to use hostnames and paths to determine where traffic should go, making it a layer 7 load balancer. For example, we can have multiple servers that are optimized for different things. If one of our servers is optimized for video, then all video requests should be routed to that specific server based on the incoming URL request.

We will cover the following recipes in this chapter:

- Creating a new application gateway
- Configuring the backend pools
- Configuring HTTP settings
- Configuring listeners
- Configuring rules
- Configuring probes
- Configuring a **Web Application Firewall (WAF)**
- Customizing WAF rules
- Creating a WAF policy

## Technical requirements

For this chapter, an Azure subscription is required.

## Creating a new application gateway

Azure Application Gateway can be used as a simple load balancer to perform traffic distribution from the frontend to the backend based on protocols and ports. But it can also expand on that and perform additional routing based on hostnames and paths. This allows us to have resource pools based on rules and also allows us to optimize performance. Using these options and performing routing based on context will increase application performance, along with providing high availability. Of course, in this case, we need to have multiple resources for each performance type in each backend pool (each performance type requests a separate backend pool).

### Getting ready

Before you start, open the browser and go to the Azure portal at https://portal.azure.com.

## How to do it...

In order to create a new application gateway, we must do the following:

1. In the Azure portal, select **Create a resource** and choose **Application Gateway** under **Networking** (or search for `application gateway` in the search bar).

2. In the new pane, we must provide information for **Subscription**, **Resource group**, **Name**, **Region**, **Tier**, **Autoscaling**, **Instance count**, **Availability zone**, and **HTTP2**. We must also select the **Virtual network** and **Subnet** that will be associated with our application gateway. You will be limited to virtual networks that are located in the region that is selected for the application gateway:

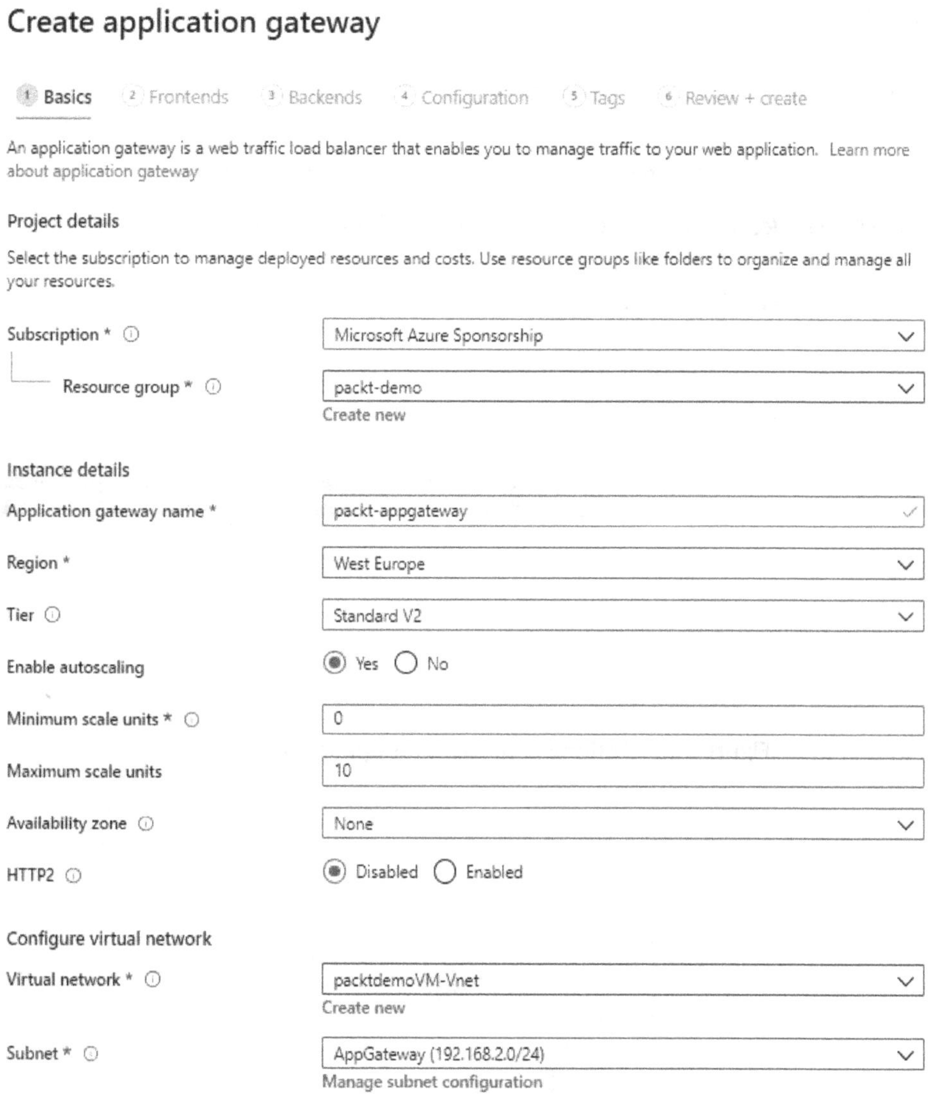

### Create application gateway

① **Basics**   ② Frontends   ③ Backends   ④ Configuration   ⑤ Tags   ⑥ Review + create

An application gateway is a web traffic load balancer that enables you to manage traffic to your web application. Learn more about application gateway

**Project details**

Select the subscription to manage deployed resources and costs. Use resource groups like folders to organize and manage all your resources.

| | |
|---|---|
| Subscription * ⓘ | Microsoft Azure Sponsorship ⌄ |
| Resource group * ⓘ | packt-demo ⌄ |
| | Create new |

**Instance details**

| | |
|---|---|
| Application gateway name * | packt-appgateway ✓ |
| Region * | West Europe ⌄ |
| Tier ⓘ | Standard V2 ⌄ |
| Enable autoscaling | ◉ Yes ◯ No |
| Minimum scale units * ⓘ | 0 |
| Maximum scale units | 10 |
| Availability zone ⓘ | None ⌄ |
| HTTP2 ⓘ | ◉ Disabled ◯ Enabled |

**Configure virtual network**

| | |
|---|---|
| Virtual network * ⓘ | packtdemoVM-Vnet ⌄ |
| | Create new |
| Subnet * ⓘ | AppGateway (192.168.2.0/24) ⌄ |
| | Manage subnet configuration |

Figure 12.1: Configuring project details for application gateway

3. Now, we fill in the **Frontends** tab. Here, we need to select the type of IP address that the frontend will use (**Public**, **Private**, or **Both**) and provide an IP (select an existing one or create a new one):

## Create application gateway

✓ Basics    ② **Frontends**    ③ Backends    ④ Configuration    ⑤ Tags    ⑥ Review + create

Traffic enters the application gateway via its frontend IP address(es). An application gateway can use a public IP address, private IP address, or one of each type.

| Frontend IP address type ⓘ | ⦿ Public    ○ Private    ○ Both |
|---|---|
| Public IP address * | (New) AppGateway-IP    ⌄ |
| | Add new |

**Figure 12.2: Selecting the Frontend IP address type**

4. Next is the **Backends** tab. We need to select **Add a backend pool**:

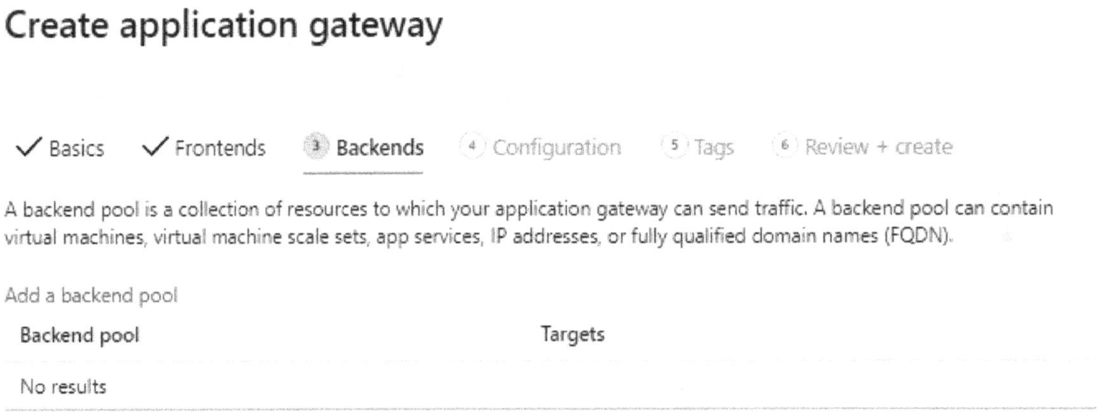

## Create application gateway

✓ Basics    ✓ Frontends    ③ **Backends**    ④ Configuration    ⑤ Tags    ⑥ Review + create

A backend pool is a collection of resources to which your application gateway can send traffic. A backend pool can contain virtual machines, virtual machine scale sets, app services, IP addresses, or fully qualified domain names (FQDN).

Add a backend pool

| Backend pool | Targets |
|---|---|
| No results | |

**Figure 12.3: Defining backends for application gateway**

5.  At this point, a new pane will open. We need to provide information for **Name** and choose whether we want to add a backend pool with or without targets. If we choose to add targets at this stage, first, we need to select **Target type**. The available types are virtual machines, virtual machine scale sets, app services, and IP addresses/FQDNs. Based on the type selection, you can add appropriate targets:

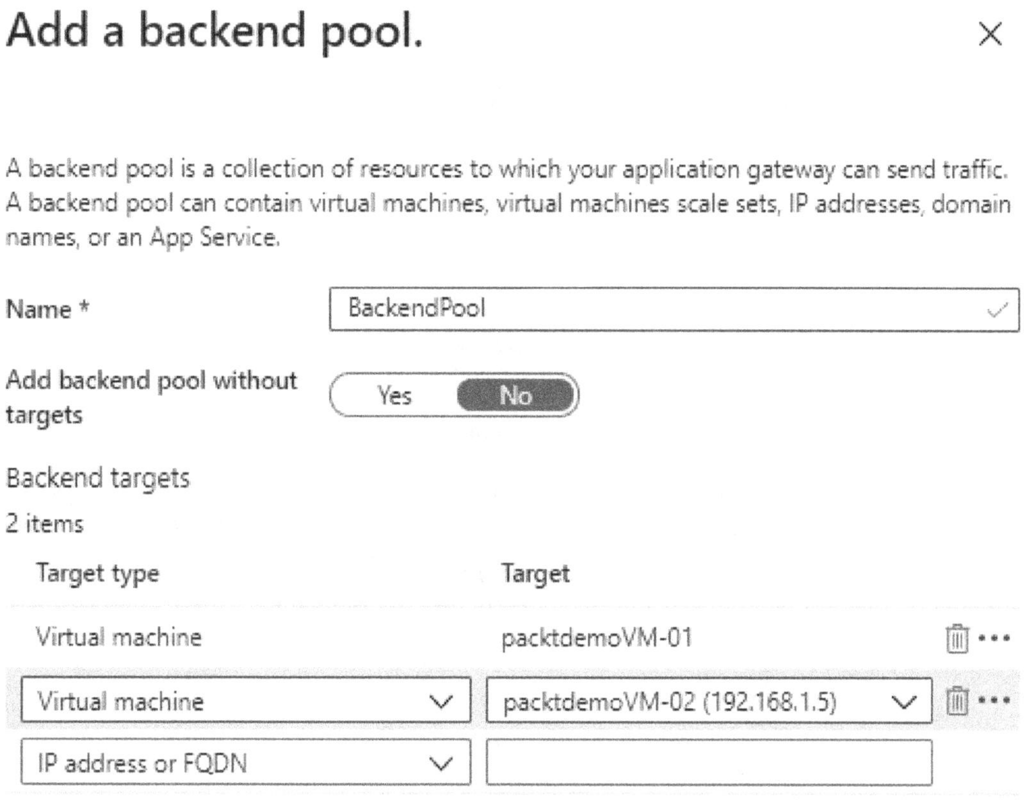

# Add a backend pool.                                        ✕

A backend pool is a collection of resources to which your application gateway can send traffic. A backend pool can contain virtual machines, virtual machines scale sets, IP addresses, domain names, or an App Service.

Name *                          | BackendPool                                    ✓ |

Add backend pool without        ( Yes    **No** )
targets

Backend targets

2 items

| Target type | | Target | |
| --- | --- | --- | --- |
| Virtual machine | | packtdemoVM-01 | 🗑 ••• |
| Virtual machine ∨ | | packtdemoVM-02 (192.168.1.5) ∨ | 🗑 ••• |
| IP address or FQDN ∨ | | | |

Figure 12.4: Adding a backend pool

6. After we have added a backend pool, we can see related information and proceed. Note that we can add more than one backend pool:

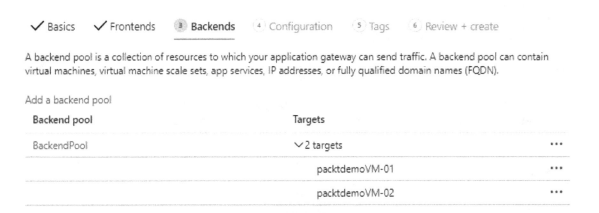

Figure 12.5: Reviewing the configuration for the backend pool

7. In the **Configuration** pane, we can see that the frontends and backends pools are in place, but we are missing a routing rule. This is mandatory in order to proceed, so we must create one by selecting **Add a routing rule**:

Figure 12.6: Creating a routing rule

8. In the new pane, we must first define a listener. For the listener, we must provide a name, select the **Frontend IP** configuration, and provide a **Port** and **Protocol** that will be monitored. We can also change the **Listener type** radio button and add a redirect URL page for errors (this can only be an Azure storage account URL):

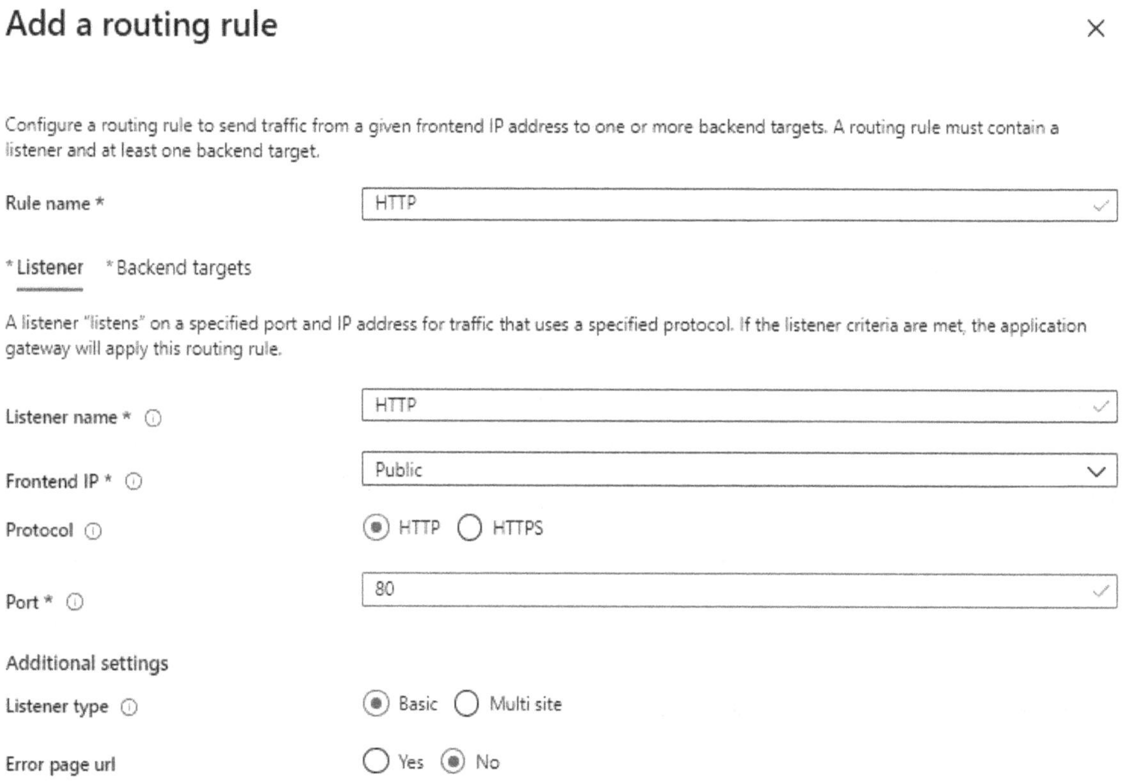

## Add a routing rule       ✕

Configure a routing rule to send traffic from a given frontend IP address to one or more backend targets. A routing rule must contain a listener and at least one backend target.

Rule name *

> HTTP ✓

\* Listener    \*Backend targets

A listener "listens" on a specified port and IP address for traffic that uses a specified protocol. If the listener criteria are met, the application gateway will apply this routing rule.

Listener name * ⓘ

> HTTP ✓

Frontend IP * ⓘ

> Public ⌄

Protocol ⓘ

   ◉ HTTP    ○ HTTPS

Port * ⓘ

> 80 ✓

Additional settings

Listener type ⓘ    ◉ Basic    ○ Multi site

Error page url    ○ Yes    ◉ No

**Figure 12.7: Configuring the listener settings for the routing rule**

9. For the routing rule, we need to configure **Backend targets** as well. In this section, we need to set **Target type**, **Backend target**, and **HTTP settings**. At this stage, we are still lacking an HTTP setting, so we need to select **Add new** under the **HTTP settings** field:

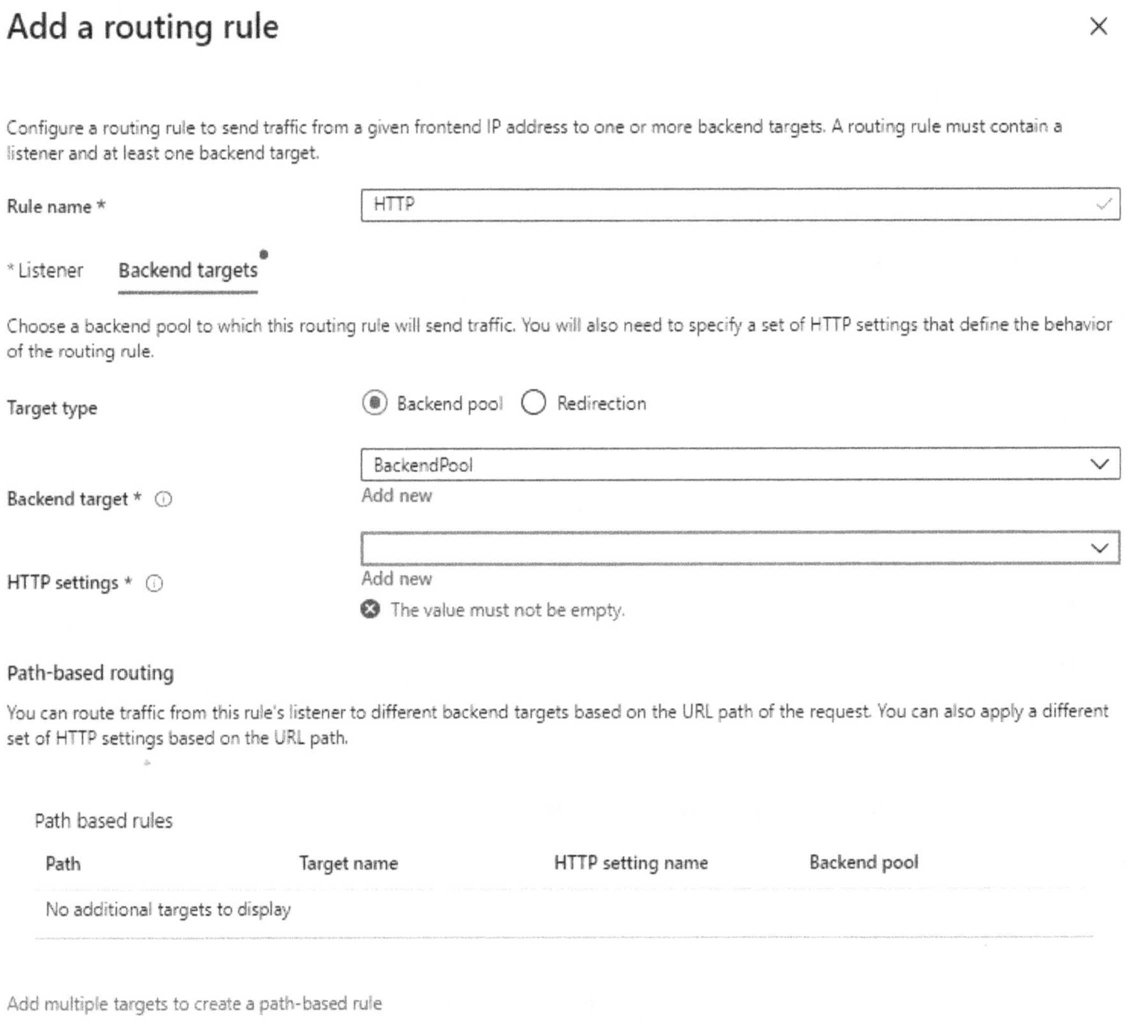

## Add a routing rule                                              ✕

Configure a routing rule to send traffic from a given frontend IP address to one or more backend targets. A routing rule must contain a listener and at least one backend target.

Rule name *                     | HTTP                                                    ✓ |

\* Listener     **Backend targets** •

Choose a backend pool to which this routing rule will send traffic. You will also need to specify a set of HTTP settings that define the behavior of the routing rule.

Target type                     ⦿ Backend pool   ◯ Redirection

| BackendPool                                             ⌄ |
Backend target * ⓘ              Add new

|                                                         ⌄ |
HTTP settings * ⓘ               Add new
                                ✕ The value must not be empty.

### Path-based routing

You can route traffic from this rule's listener to different backend targets based on the URL path of the request. You can also apply a different set of HTTP settings based on the URL path.

Path based rules

| Path | Target name | HTTP setting name | Backend pool |
| --- | --- | --- | --- |
| No additional targets to display | | | |

Add multiple targets to create a path-based rule

**Figure 12.8: Configuring backend targets for the routing rule**

10. In the new pane, first, we need to provide our HTTP setting with a name and add details for **Backend protocol** and **Backend port**. We must also enable or disable **Cookie-based affinity** and **Connection draining** before specifying the **Request time-out (seconds)** period. We can enable or disable the **Create custom probes** and **Override with new host name** settings:

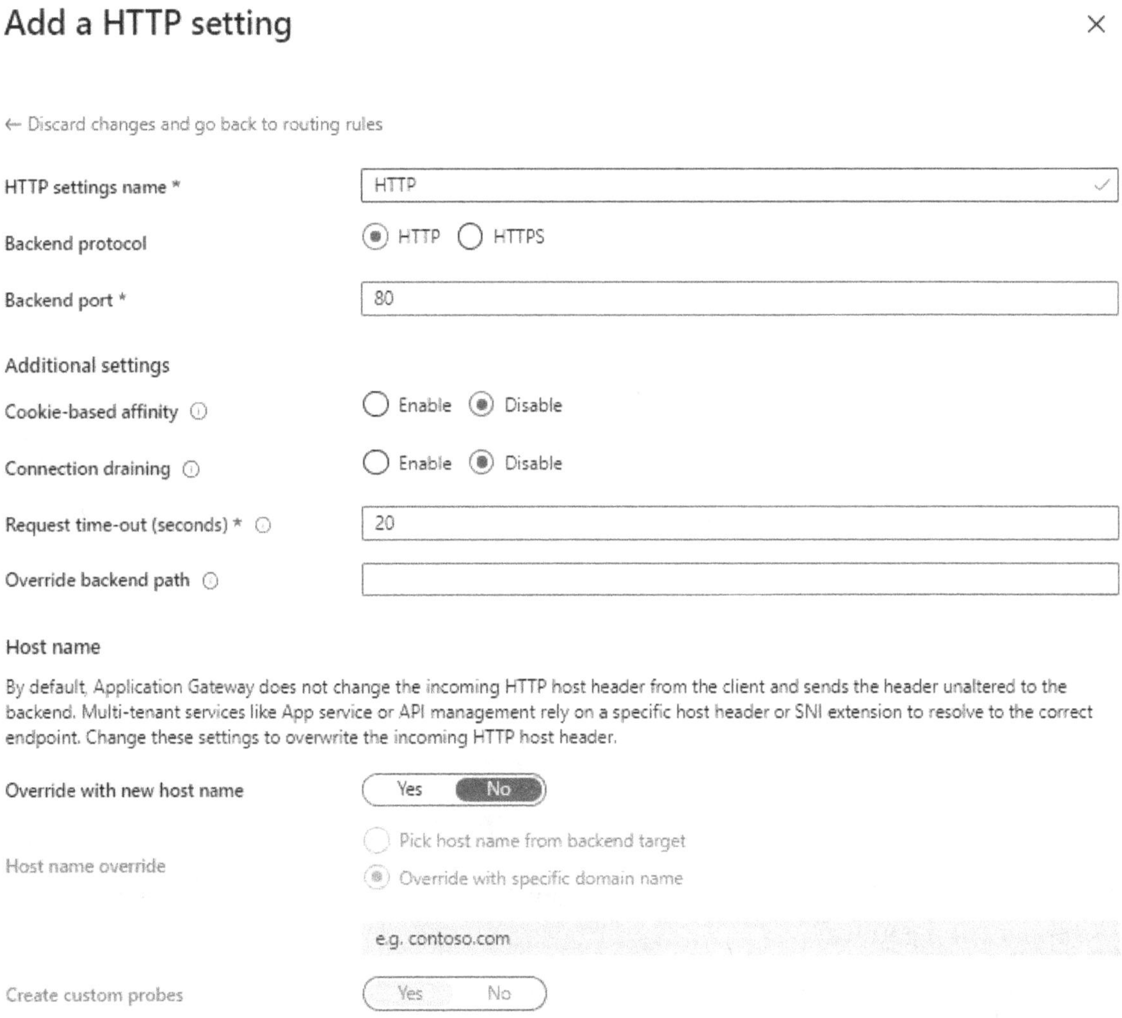

## Add a HTTP setting                                              ×

← Discard changes and go back to routing rules

| | |
|---|---|
| HTTP settings name * | HTTP |
| Backend protocol | ⦿ HTTP  ◯ HTTPS |
| Backend port * | 80 |

Additional settings

| | |
|---|---|
| Cookie-based affinity ⓘ | ◯ Enable  ⦿ Disable |
| Connection draining ⓘ | ◯ Enable  ⦿ Disable |
| Request time-out (seconds) * ⓘ | 20 |
| Override backend path ⓘ | |

Host name

By default, Application Gateway does not change the incoming HTTP host header from the client and sends the header unaltered to the backend. Multi-tenant services like App service or API management rely on a specific host header or SNI extension to resolve to the correct endpoint. Change these settings to overwrite the incoming HTTP host header.

| | |
|---|---|
| Override with new host name | Yes  **No** |
| Host name override | ◯ Pick host name from backend target<br>⦿ Override with specific domain name |
| | e.g. contoso.com |
| Create custom probes | Yes  No |

**Figure 12.9: Adding an HTTP setting**

11. After the HTTP setting is created, it will be automatically added to our routing rule, which we can now finish:

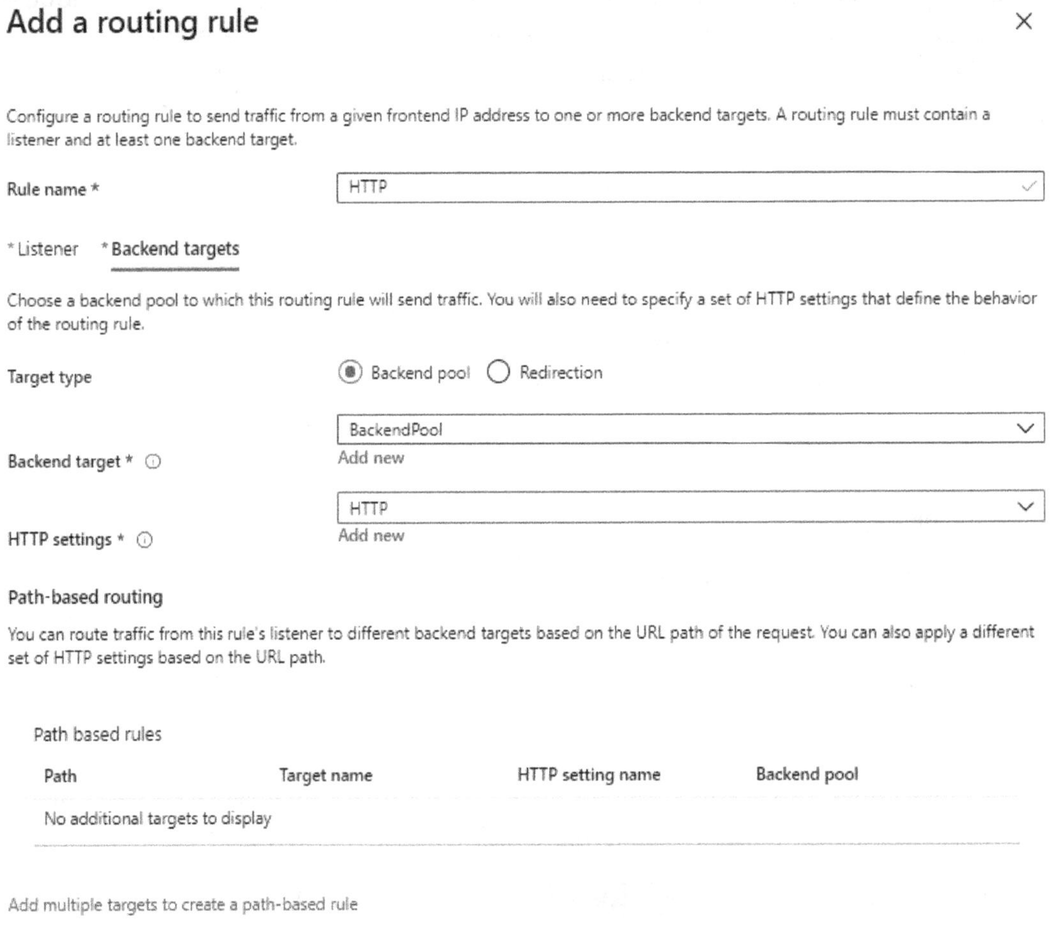

## Add a routing rule                                                    ✕

Configure a routing rule to send traffic from a given frontend IP address to one or more backend targets. A routing rule must contain a listener and at least one backend target.

Rule name *                        | HTTP                                              ✓ |

*Listener    *Backend targets

Choose a backend pool to which this routing rule will send traffic. You will also need to specify a set of HTTP settings that define the behavior of the routing rule.

Target type                        ◉ Backend pool   ○ Redirection

Backend target * ⓘ                 | BackendPool                                       ⌄ |
                                   Add new

HTTP settings * ⓘ                  | HTTP                                              ⌄ |
                                   Add new

### Path-based routing

You can route traffic from this rule's listener to different backend targets based on the URL path of the request. You can also apply a different set of HTTP settings based on the URL path.

Path based rules

| Path | Target name | HTTP setting name | Backend pool |
| --- | --- | --- | --- |
| No additional targets to display | | | |

Add multiple targets to create a path-based rule

Figure 12.10: Final configuration for adding a routing rule

12. The configuration is now complete, and we can go ahead and deploy our application gateway:

## Create application gateway

✓ Basics   ✓ Frontends   ✓ Backends   ① **Configuration**   5 Tags   6 Review + create

Create routing rules that link your frontend(s) and backend(s). You can also add more backend pools, add a second frontend IP configuration if you haven't already, or edit previous configurations.

| **Frontends** | **Routing rules** | **Backend pools** |
| --- | --- | --- |
| + Add a frontend IP | + Add a routing rule | + Add a backend pool |
| Public: (new) AppGateway-IP   🗑 ··· | HTTP   🗑 ···<br>Manage HTTP settings | BackendPool   🗑 ··· |

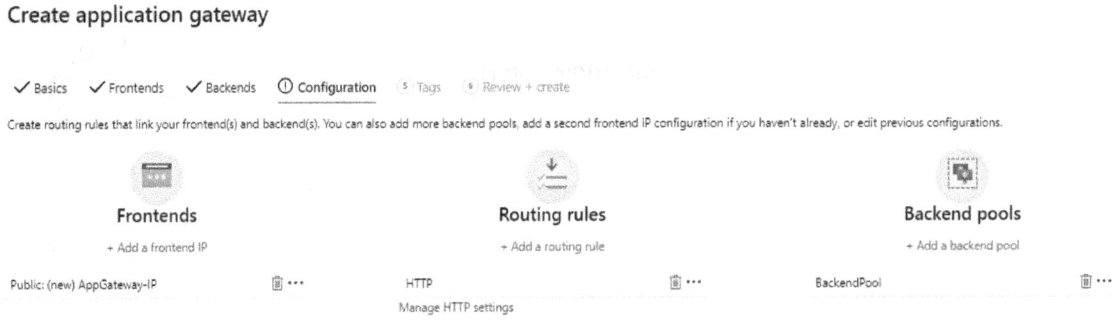

Figure 12.11: Deploying our application gateway

## How it works...

Azure Application Gateway is very similar to Azure Load Balancer, with some additional options. It will route traffic coming to the front end of the application gateway to a defined backend based on rules that we define. In addition to routing based on protocols and ports, the application gateway also allows defined routing based on paths and protocols. Using these additional rules, we can route incoming requests to endpoints that are optimized for certain roles. For example, we can have multiple backend pools with different settings that are optimized to perform only specific tasks. Based on the nature of the incoming requests, the application gateway will route the requests to the appropriate backend pool. This approach, along with high availability, will provide better performance by routing each request to a backend pool that will process the request in a more optimized way.

We can set up autoscaling for application gateway (available only for V2) with additional information for the minimum and maximum number of units. This way, application gateway will scale based on demand and ensure that performance is not impacted, even with the maximum number of requests.

## Configuring the backend pools

After the application gateway is created, we must define the backend pools. Traffic coming to the front end of the application gateway will be forwarded to the backend pools. Backend pools in application gateways are the same as backend pools in load balancers and are defined as possible destinations where traffic will be routed based on other settings that will be added in future recipes in this chapter.

### Getting ready

Before you start, open the browser and go to the Azure portal at https://portal.azure. com.

## How to do it...

In order to add backend pools to our application gateway, we must do the following:

1.  In the Azure portal, locate the previously created application gateway.

2.  In the **Application gateway** pane, under **Settings**, select **Backend pools**. Select **Add** to add a new backend pool or select an existing one to edit:

**Figure 12.12: Adding a backend pool to our application gateway**

3.  In the new pane, the only difference between new and existing pools is the name. For a new pool, we must provide the name of the backend pool, and for existing pools, this option is grayed out and cannot be edited. For both new and existing pools, we must provide the type of target. The available types are virtual machines, virtual machine scale sets, app services, and IP addresses/FQDNs. Based on the type selection, you can add appropriate targets:

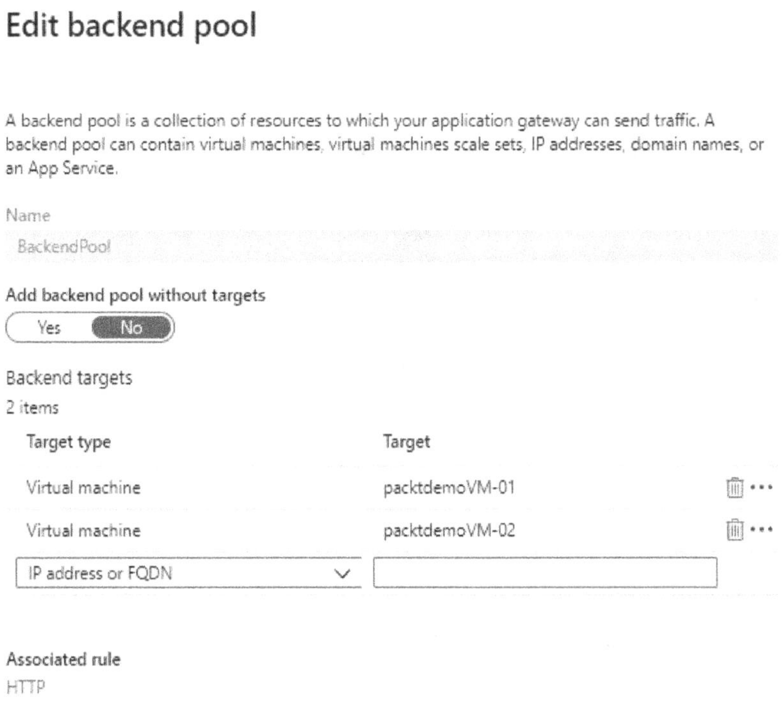

## Edit backend pool

A backend pool is a collection of resources to which your application gateway can send traffic. A backend pool can contain virtual machines, virtual machines scale sets, IP addresses, domain names, or an App Service.

Name

BackendPool

Add backend pool without targets

Yes      No

Backend targets

2 items

| Target type | Target | | |
|---|---|---|---|
| Virtual machine | packtdemoVM-01 | 🗑 | ••• |
| Virtual machine | packtdemoVM-02 | 🗑 | ••• |
| IP address or FQDN ∨ | | | |

Associated rule

HTTP

**Figure 12.13: Providing the target type for the backend pool**

## How it works...

With backend pools, we define targets to which traffic will be forwarded. As the application gateway allows us to define routing for each request, it's best to have targets based on performance and types grouped in the same way. For example, if we have multiple web servers, these should be placed in the same backend pool. Servers used for data processing should be placed in a separate pool, and servers used for video in another separate pool. This way, we can separate pools based on performance types, and route traffic based on operations that need to be completed.

This will increase the performance of our application, as each request will be processed by the resource best suited for a specific task. To achieve high availability, we should add more servers to each backend pool.

# Configuring HTTP settings

HTTP settings in application gateways are used for validation and various traffic settings. Their main purpose is to ensure that requests are directed to the appropriate backend pool. Some other HTTP settings are also included, such as affinity or connection draining. Override settings are also part of HTTP settings–these will allow you to redirect if an incomplete or incorrect request is sent.

## Getting ready

Before you start, open the browser and go to the Azure portal at https://portal.azure.com.

## How to do it...

In order to add HTTP settings to our application gateway, we must do the following:

1. In the Azure portal, locate the previously created application gateway.

2. In the **Application gateway** pane, under **Settings**, select **HTTP settings**. Select **Add** to add a new HTTP setting or select an existing one to edit:

Figure 12.14: Locating HTTP settings in the Application gateway pane

3.  In the new pane, first, we need to provide a name (if you are editing an existing HTTP setting, this option will be grayed out). The next options allow us to disable or enable **Cookie-based affinity** and **Connection draining**. Further to this, we select our **Protocol**, **Port**, and the **Request time-out (seconds)** period. Optional settings allow us to configure **Use custom probe** and **Override with new host name**:

## Add HTTP setting

HTTP settings name

HTTP

Backend protocol
◉ HTTP   ○ HTTPS

Backend port *

80

Additional settings

Cookie-based affinity ⓘ
○ Enable   ◉ Disable

Connection draining ⓘ
○ Enable   ◉ Disable

Request time-out (seconds) * ⓘ

20

Override backend path ⓘ

Host name

By default, Application Gateway does not change the incoming HTTP host header from the client and sends the header unaltered to the backend. Multi-tenant services like App service or API management rely on a specific host header or SNI extension to resolve to the correct endpoint. Change these settings to overwrite the incoming HTTP host header.

Override with new host name
( Yes    No )

Host name override
○ Pick host name from backend target
◉ Override with specific domain name

e.g. contoso.com

Use custom probe ⓘ
○ Yes ◉ No

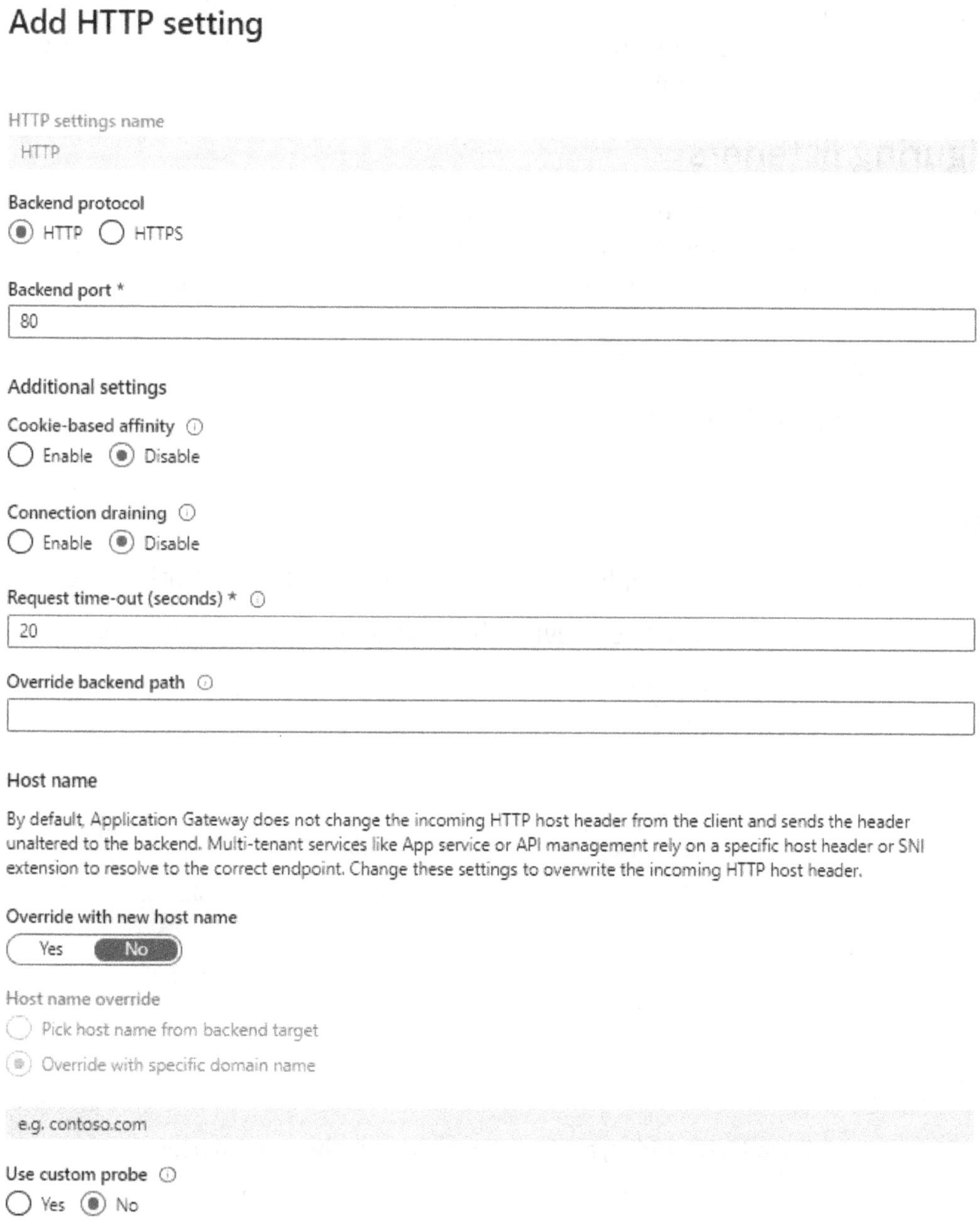

Figure 12.15: Configuring HTTP settings

## How it works...

As previously mentioned, the main purpose of HTTP settings is to ensure that requests are directed to the correct backend pool. However, various other options are available. Cookie-based affinity allows us to route requests from the same source to the same target server in the backend pool. Connection draining will control the behavior when the server is removed from the backend pool. If this is enabled, the server will help maintain in-flight requests to the same server. Override settings allow us to override the path of the URL to a different path or a completely new domain, before forwarding the request to the backend pool.

# Configuring listeners

**Listeners** in an application gateway listen for any incoming requests. After a new request is detected, it's forwarded to the backend pool based on the rules and settings we have defined. In this recipe, we will add a new listener to our application gateway.

## Getting ready

Before you start, open the browser and go to the Azure portal at https://portal.azure.com.

## How to do it...

In order to add a listener to an application gateway, we must do the following:

1.  In the Azure portal, locate the previously created application gateway.

2.  In the **Application gateway** pane, under **Settings**, select **Listeners**, then select **Add listener** to add a new listener, or edit an existing one:

Figure 12.16: Adding a new listener through the Azure portal

3. In the new pane, we need to provide a name for the listener (if you are editing an existing listener, this option will be grayed out), select the **Frontend IP** configuration, and provide the **Port** and **Protocol** that will be monitored. Additionally, we can set up the **Listener type** and a custom URL page for errors:

## HTTP

packt-appgateway

Listener name ⓘ

HTTP

Frontend IP * ○

Public ∨

Port * ⓘ

80 ✓

Protocol ⓘ
⦿ HTTP ○ HTTPS

Associated rule
HTTP

Additional settings

Listener type ⓘ
⦿ Basic ○ Multi site

Error page url
○ Yes ⦿ No

**Figure 12.17: Configuring the listener settings for our application gateway**

## How it works...

A listener monitors for new requests coming to the application gateway. Each listener monitors only one frontend IP address and only one port. If we have two frontend IPs (one public and one private) and traffic coming in over multiple protocols and ports, we must create a listener for each IP address and each port that traffic may be coming to.

The basic type of listener is used when the listener listens to a single domain; it's usually used when we host a single application behind an application gateway. A multi-site listener is used when we have more than one application behind the application gateway and we need to configure routing based on a host name or domain name.

# Configuring rules

Rules in application gateways are used to determine how traffic flows. Different settings determine where a specific request is forwarded to and how this is done.

## Getting ready

Before you start, open the browser and go to the Azure portal at https://portal.azure.com.

## How to do it...

In order to add a rule to the application gateway, we must do the following:

1. In the Azure portal, locate the previously created application gateway.

2. In the **Application gateway** pane, under **Settings**, select **Rules**. Add a new rule or select an existing one to edit:

**Figure 12.18: Adding a routing rule for our application gateway**

3.  In the new pane, we must provide a name for the new rule (if you are editing an existing rule, this option is grayed out) and select the **Listener**, as shown in *Figure 12.19*:

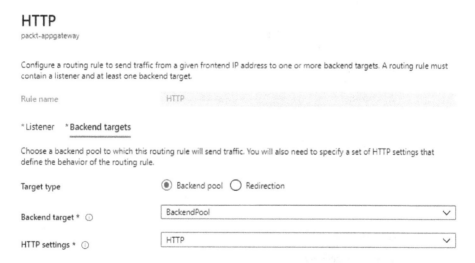

## HTTP
packt-appgateway

Configure a routing rule to send traffic from a given frontend IP address to one or more backend targets. A routing rule must contain a listener and at least one backend target.

Rule name                   HTTP

\*Listener    \*Backend targets

A listener "listens" on a specified port and IP address for traffic that uses a specified protocol. If the listener criteria are met, the application gateway will apply this routing rule.

Listener \*                 HTTP                                              ∨

Figure 12.19: Configuring the routing rule

4.  We also need to set up a backend target, where we need to define **Target type** and select options for **Backend target** and **HTTP settings**:

## HTTP
packt-appgateway

Configure a routing rule to send traffic from a given frontend IP address to one or more backend targets. A routing rule must contain a listener and at least one backend target.

Rule name                   HTTP

\*Listener    \*Backend targets

Choose a backend pool to which this routing rule will send traffic. You will also need to specify a set of HTTP settings that define the behavior of the routing rule.

Target type                 ⦿ Backend pool  ○ Redirection

Backend target \* ⓘ          BackendPool                                       ∨

HTTP settings \* ⓘ           HTTP                                              ∨

Figure 12.20: Setting up a backend target for our routing rule

## How it works...

Using rules, we can tie some previously created settings together. We define a listener that specifies which request on what IP address we are expecting on which port. Then, these requests are forwarded to the backend pool; forwarding is performed based on the HTTP settings. Optionally, we can also add redirection to the rules.

# Configuring probes

Probes in application gateway are used to monitor the health of the backend targets. Each endpoint is monitored, and if one is found to be unhealthy, it is temporarily taken out of rotation and requests are not forwarded. Once the status changes, it's added back. This prevents requests from being sent to unhealthy endpoints that can't serve the request.

## Getting ready

Before you start, open the browser and go to the Azure portal at https://portal.azure.com.

## How to do it...

In order to add a probe to our application gateway, we must do the following:

1. In the Azure portal, locate the previously created application gateway.

2. In the **Application gateway** pane, under **Settings**, select **Health probes**. Select **Add** to add the new probe:

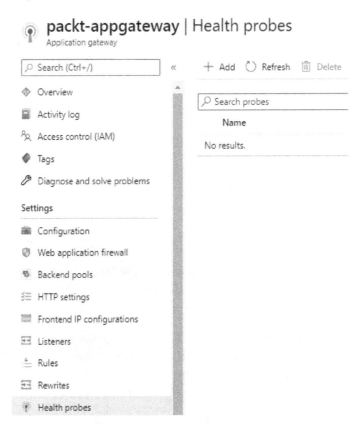

Figure 12.21: Adding a new health probe

3.  In the new pane, we must provide the **Name** of the probe (this option will be grayed out if an existing probe is edited), along with the **Protocol**, **Host**, and **Path**. We also need to set the **Interval (seconds)**, **Timeout (seconds)**, and **Unhealthy threshold** settings. We can also choose to configure **Use probe matching conditions** and associate **HTTP settings**:

Figure 12.22: Configuring the health probe details

## How it works...

**Protocol**, **Host**, and **Path** define what probe is being monitored. **Interval** defines how often checks are performed. **Timeout** defines how much time must pass before the check is declared to have failed. Finally, **Unhealthy threshold** is used to set how many failed checks must occur before the endpoint is declared unavailable.

# Configuring a Web Application Firewall (WAF)

WAF is an additional setting for the application gateway. It's used to increase the security of applications behind the application gateway, and it also provides centralized protection.

## Getting ready

To enable a WAF, we must set the application gateway to the WAF tier. To do so, we must do the following:

1. In the **Application gateway** pane, go to **Web application firewall**, under **Settings**. Change the **Tier** selection from **Standard V2** to **WAF V2** and select **Save**:

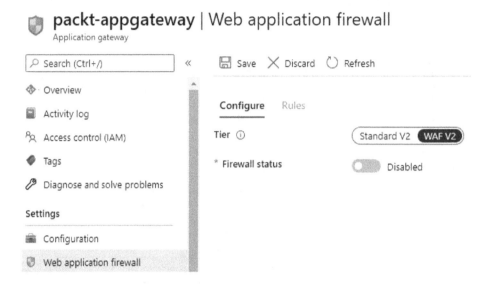

Figure 12.23: Setting the application gateway to the WAF V2 tier

## How to do it...

After the application gateway is set to WAF, we can enable and set the firewall rules. To do so, we must do the following:

1. In the **Application gateway** pane, go to **Web application firewall**, under **Settings**, and enable **Firewall status**. After we set **Firewall status** to **Enabled**, a new set of options will appear:

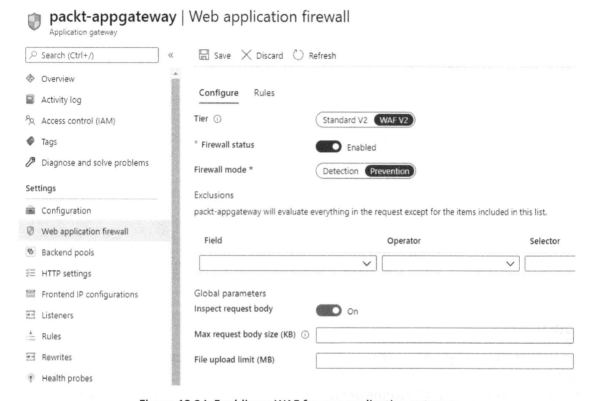

Figure 12.24: Enabling a WAF for our application gateway

2. We must select a **Firewall mode**, set an exclusion list, and specify the **Global parameters** as follows:

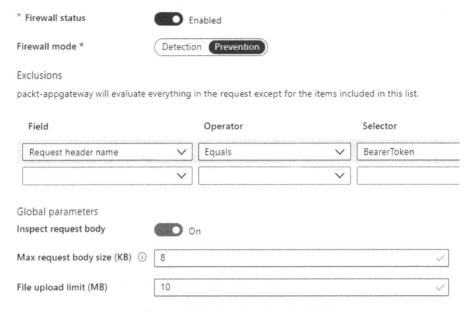

**Figure 12.25: Configuring the WAF**

## How it works...

The WAF feature helps increase security by checking all incoming traffic. As this can slow down performance, we can exclude some items that are creating false positives, especially when it comes to items of significant size. Excluded items will not be inspected. A WAF can work in two modes: detection and prevention. Detection will only detect if a malicious request is sent, while prevention will stop any such request.

# Customizing WAF rules

A WAF comes with a predetermined set of rules. These rules are enforced to increase application security and prevent malicious requests. We can change these rules to address specific issues or requirements as needed.

## Getting ready

Before you start, open the browser and go to the Azure portal at https://portal.azure. com.

## How to do it...

In order to change the WAF rules, we must do the following:

1. Select **Web application firewall** under **Settings** in the **Application gateway** pane.

2. Select **Rules** in the WAF settings. Select **Enabled** under **Advanced rule configuration**, as shown in *Figure 12.26*:

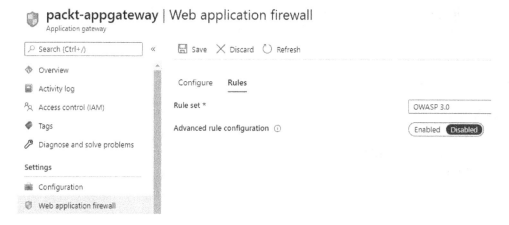

Figure 12.26: Enabling Advanced rule configuration

3. The rules will appear in the form of a list. We can check or uncheck boxes to enable or disable rules:

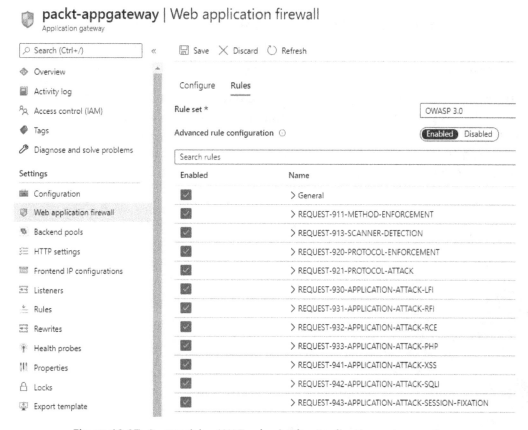

Figure 12.27: Customizing WAF rules in the Application gateway pane

## How it works...

A WAF comes with all rules activated by default. This can slow down performance, so we can disable some of the rules if needed. Also, there are three rule sets available— **OWASP 2.2.9**, **OWASP 3.0**, and **OWASP 3.1**. The default (and recommended) rule set is **OWASP 3.0**, but we can switch between rule sets as required.

# Creating a WAF policy

A WAF policy allows us to handle WAF settings and configurations as a separate resource. By doing so, we can apply the same policy to multiple resources instead of individual application gateways. A WAF policy can be associated with Application Gateway, Front Door, or CDN.

## Getting ready

Before you start, open the browser and go to the Azure portal at https://portal.azure.com.

## How to do it...

In order to create a new application gateway, we must do the following:

1.  In the Azure portal, select **Create a resource** and choose **Web Application Firewall** under **Networking** (or search for `Web Application Firewall` in the search bar).

2.  In the new pane, we must complete the **Basics** section first. We need to set what the policy is going to be used for (Application Gateway, Front Door, or CDN), configure **Subscription** and **Resource group**, and fill in the **Policy name** and **Location** fields. Additionally, we can set whether the policy will be enabled or disabled once it's created:

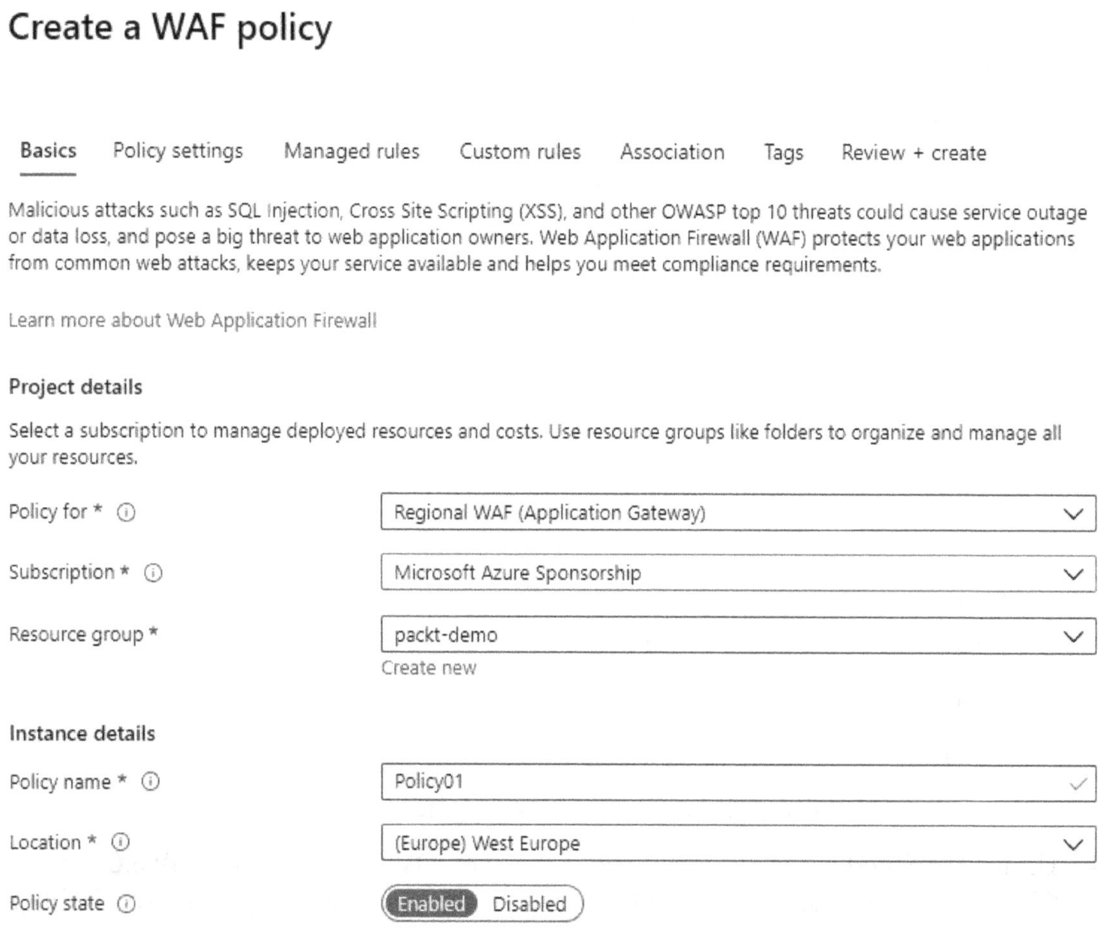

# Create a WAF policy

Basics    Policy settings    Managed rules    Custom rules    Association    Tags    Review + create

Malicious attacks such as SQL Injection, Cross Site Scripting (XSS), and other OWASP top 10 threats could cause service outage or data loss, and pose a big threat to web application owners. Web Application Firewall (WAF) protects your web applications from common web attacks, keeps your service available and helps you meet compliance requirements.

Learn more about Web Application Firewall

## Project details

Select a subscription to manage deployed resources and costs. Use resource groups like folders to organize and manage all your resources.

| | |
|---|---|
| Policy for * ⓘ | Regional WAF (Application Gateway) ⌄ |
| Subscription * ⓘ | Microsoft Azure Sponsorship ⌄ |
| Resource group * | packt-demo ⌄ |
| | Create new |

## Instance details

| | |
|---|---|
| Policy name * ⓘ | Policy01 ✓ |
| Location * ⓘ | (Europe) West Europe ⌄ |
| Policy state ⓘ | ( Enabled  Disabled ) |

Figure 12.28: Creating a new WAF policy

3. In **Policy settings**, we can set **Mode** to **Detection** or **Prevention**, along with **Exclusions** and **Global parameters**:

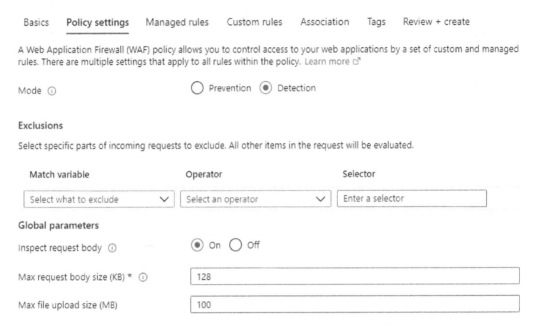

Figure 12.29: Configuring policy settings for your WAF policy

4. Under **Managed rules**, we can select a rule set (**OWASP 2.2.9**, **OWASP 3.0**, or **OWASP 3.1**) and disable some rules if needed (it is not recommended to disable rules unless necessary):

Figure 12.30: Setting rules for your WAF policy

5. Under **Custom rules**, we can add additional rules if needed. Select **Add custom rule** to add one:

## Create a WAF policy

Basics    Policy settings    Managed rules    **Custom rules**    Association    Tags    Review + create

Configure a policy with custom-authored rules. Once a rule is matched, the corresponding action defined in the rule is applied to the request. Once such a match is processed, rules with lower priorities are not processed further. A smaller integer value for a rule denotes a higher priority. Learn more ◻

➕ Add custom rule

| Priority | Name | Action |
|----------|------|--------|
| No custom rules to display. | | |

Figure 12.31: Adding a custom rule to our WAF policy

6. This will open a new pane that will allow us to define a custom rule. We need to fill in the **Custom rule name** field and set **Priority** to **1**. Under **Conditions**, we are creating a match type and variables that need to be matched in order to trigger the rule. Finally, we set a response (allow, deny, or log):

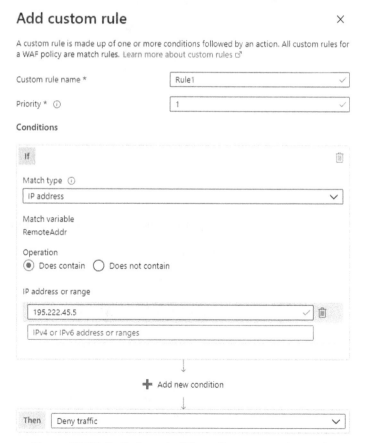

## Add custom rule                                                    ✕

A custom rule is made up of one or more conditions followed by an action. All custom rules for a WAF policy are match rules. Learn more about custom rules ◻

Custom rule name *                    Rule1                              ✓

Priority * ⓘ                          1                                  ⌄

Conditions

If                                                                       🗑

Match type ⓘ
IP address                                                              ⌄

Match variable
RemoteAddr

Operation
◉ Does contain    ◯ Does not contain

IP address or range
195.222.45.5                                          ✓  🗑
IPv4 or IPv6 address or ranges

➕ Add new condition

Then    Deny traffic                                                    ⌄

Figure 12.32: Defining conditions for your custom rule

7.  Once the custom rule is created, it will appear in the list and we can proceed to the **Association** section:

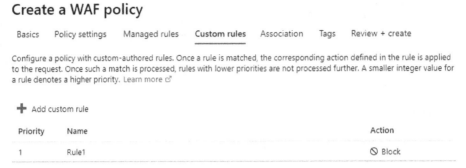

Figure 12.33: List showing the new custom rule

8.  In the **Association** section, we are creating an association with the service we want to apply the policy to. This section will depend on the previously selected service type (in our case, application gateway). Select **Associate an application gateway**:

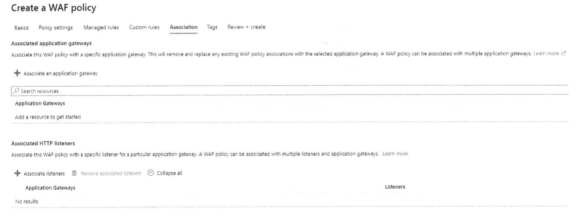

Figure 12.34: Creating an association with our application gateway

9.  In the new pane, select **Application gateway** from the drop-down menu. Note that only **WAF V2 SKU** is supported:

Figure 12.35: Choosing your application gateway from the drop-down menu

10. Once **Application gateway** is selected, we need to associate listners. Select **Associate listener** under **Associate HTTP listners**. In the new pane, from the drop-down menu, select the listener you want to use:

## Associate listeners in an appli...    ✕

Application Gateway (WAF v2 SKU) * ⓘ

| packt-appgateway | ∨ |
| --- | --- |

Listeners *

| HTTP | ∨ |
| --- | --- |

Figure 12.36: Selecting the listener from the drop-down menu

11. Once the listener is associated, we can start creating our WAF policy:

### Create a WAF policy

Basics    Policy settings    Managed rules    Custom rules    **Association**    Tags    Review + create

**Associated application gateways**

Associate this WAF policy with a specific application gateway. This will remove and replace any existing WAF policy associations with the selected application gateway. A WAF policy can be associated with multiple application gateways. Learn more. ◻

✚ Associate an application gateway

🔍 Search resources

**Application Gateways**

packt-appgateway

**Associated HTTP listeners**

Associate this WAF policy with a specific listener for a particular application gateway. A WAF policy can be associated with multiple listeners and application gateways. Learn more.

✚ Associate listeners    🗑 Remove associated listeners    ⌄ Collapse all

| Application Gateways | Listeners |
| --- | --- |
| ∨ packt-appgateway | 1 |
| ☐ | HTTP |

Figure 12.37: Final configuration of our new WAF policy

## How it works...

Our WAF policy contains all the required settings and configuration for our WAF and it can be associated with Application Gateway, Front Door, or CDN. It can be associated with multiple resources but only one type at a time. The **Mode** determines what kind of action is going to be taken when an issue is detected. **Prevention** will block suspicious requests, and **Detection** will only create a log entry.

# 13

# Azure Front Door and Azure CDN

Several networking services in Microsoft Azure are dedicated to application delivery. **Azure Front Door** and **Azure CDN** are services that allow us to create applications for global delivery and take advantage of the global network of Azure datacenters. Leveraging this capability, we can provide the same experience to our users, irrespective of their physical location.

We will cover the following recipes in this chapter:

- Creating an Azure Front Door instance
- Creating an Azure CDN profile

## Technical requirements

For this chapter, the following is required:

- An Azure subscription

# Creating an Azure Front Door instance

Azure Front Door is used for the global routing of web traffic for applications distributed across different Azure regions. With Azure Front Door, we can define, manage, and monitor the routing of our web traffic and enable quick global failover. It enables us to deliver our applications with the best performance and high availability. Azure Front Door is an L7 load balancer, similar to Application Gateway. However, there is a difference as regards global distribution. In terms of global distribution, it is similar to another service–Traffic Manager. Essentially, Azure Front Door combines the best features of Application Gateway and Traffic Manager–the security of Application Gateway and the distribution capability of Traffic Manager.

## Getting ready

Azure Front Door requires services that will be added to the backend pool. You can use a script from the *Getting ready* section of the *Adding an endpoint* recipe, under *Chapter 11, Traffic Manager*.

Next, open the browser and go to the Azure portal via https://portal.azure.com.

## How to do it...

In order to create a new Azure Front Door instance, take the following steps:

1.  In the Azure portal, select **Create a resource** and choose **Front Door** under **Networking** (or search for `Front Door` in the search bar).

2.  In the new pane, we have several sections to cover. Under **Basics**, we need to provide details for **Subscription** and **Resource group**. **Resource group location** is automatically selected and grayed out:

# Create a Front Door

Basics    Configuration    Tags    Review + create

Azure Front Door Service is Microsoft's highly available and scalable web application acceleration platform and global HTTP(s) load balancer. It provides built-in DDoS protection and application layer security and caching. Front Door enables you to build applications that maximize and automate high-availability and performance for your end-users. Use Front Door with Azure services including Web/Mobile Apps, Cloud Services and Virtual Machines – or combine it with on-premises services for hybrid deployments and smooth cloud migration.  Learn more about Front Door

**PROJECT DETAILS**

Select a subscription to manage deployed resources and costs. Use resource groups like folders to organize and manage all your resources.

Figure 13.1: Providing Subscription and Resource group details

3.  In the **Configuration** section, we need to provide details for **Frontends/domains**, **Backend pools**, and **Routing rules**. Click on the **Frontends/domains** box to launch the configuration pane:

Figure 13.2: Selecting the Frontends/domains configuration option

4. In the new pane, we must provide a host name and select whether we want to enable **SESSION AFFINITY** and **WEB APPLICATION FIREWALL**:

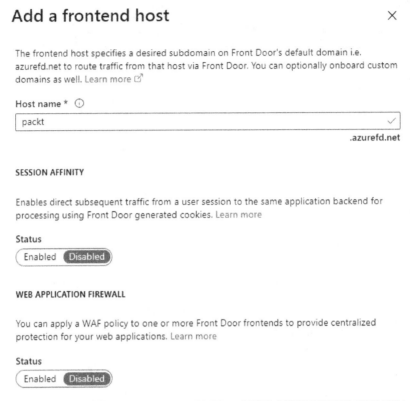

Figure 13.3: Enabling SESSION AFFINITY and WEB APPLICATION FIREWALL

5. Once the front end has been created, we are back in the **Configuration** section. Select **Backend pools** to launch the next configuration pane:

Figure 13.4: Selecting the Backend pools configuration option

6. We need to provide a name for our backend pool and add services to it. To add a backend, select the **Add a backend** option:

Figure 13.5: Adding a backend pool

7. To add a backend, we must select **Backend host type** and **Subscription** first. Based on our selection, we will be allowed to choose services (of a selected type in the selected subscription) under **Backend host name**. We also need to provide details for **Backend host header**, ports (**HTTP** and **HTTPS**), **Priority**, and **Weight**. Finally, we need to select the **Enabled** option for **Status**:

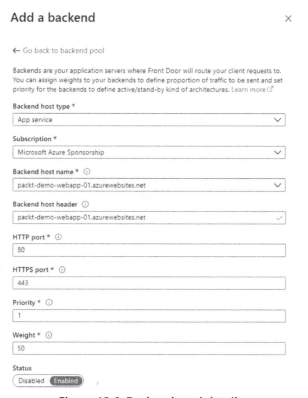

Figure 13.6: Backend pool details

8. Repeat this process to add at least one other endpoint to the backend pool:

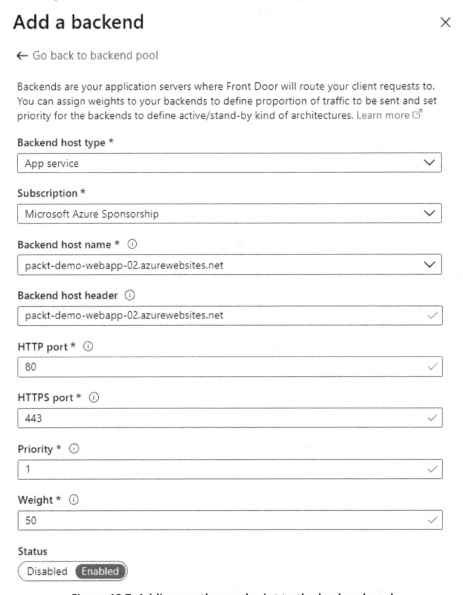

**Figure 13.7: Adding another endpoint to the backend pool**

9.  Once we have added sufficient endpoints to the backend pool, we can proceed with configuration:

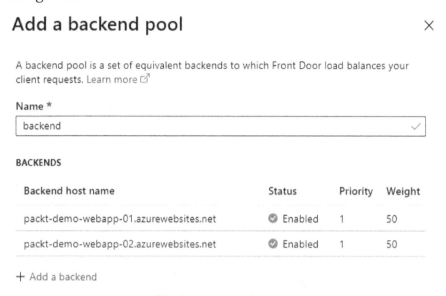

Figure 13.8: Configuring the backend pool

10. Health probes require information for **Path** (use **/** for the default option or add your own), **Protocol** (**HTTP** or **HTTPS**), **Probe method** (**HEAD** or **GET**), and **Interval** in seconds (how often the probe will check the health of the backend):

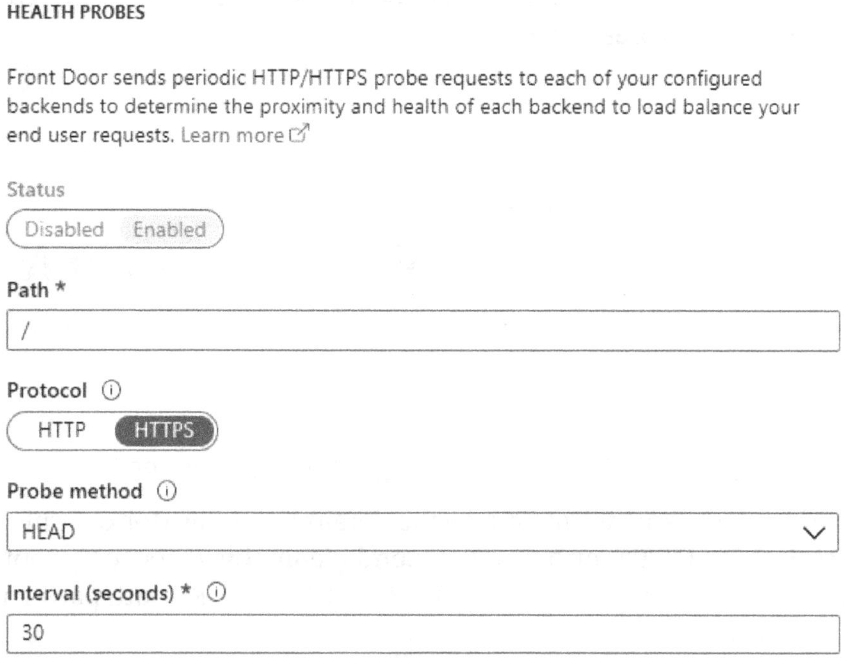

HEALTH PROBES

Front Door sends periodic HTTP/HTTPS probe requests to each of your configured backends to determine the proximity and health of each backend to load balance your end user requests. Learn more ☐

Status

( Disabled    Enabled )

Path *

/

Protocol ⓘ

( HTTP    HTTPS )

Probe method ⓘ

HEAD

Interval (seconds) * ⓘ

30

Figure 13.9: Configuring health probes to check backend health

11. Under the **LOAD BALANCING** section, we must provide information for **Sample size**, **Successful samples required**, and **Latency sensitivity**:

LOAD BALANCING

Configure the load balancing settings to define what sample set we need to use to call the backend as healthy or unhealthy. The latency sensitivity with value zero (0) means always send it to the fastest available backend, else Front Door will round robin traffic between the fastest and the next fastest backends within the configured latency sensitivity. Learn more ☐

Sample size * ⓘ

| 4 |
|---|

Successful samples required * ⓘ

| 2 |
|---|

Latency sensitivity (in milliseconds) * ⓘ

| 0 |
|---|

Figure 13.10: The LOAD BALANCING pane

12. Once we have added all the necessary information, we can create a back end pool. This will take us back to the **Configuration** section again. Select **Routing rules** to launch the **Routing rules** pane:

Figure 13.11: Selecting the Routing rules configuration option

13. In the **Add a rule** pane, we must provide details for **Name** (for our rule), **Accepted protocol** (HTTP, HTTPS, or both), **Frontends/domains** (choose the option selected previously), and **PATTERNS TO MATCH** (the URL path patterns that the route will accept):

## Add a rule                                                    ✕

A routing rule maps your frontend host and a matching URL path pattern to a specific backend pool. Learn more ☐

**Name** *

| rule1                                                        ✓ |

**Accepted protocol** ⓘ

| HTTP and HTTPS                                               ⌄ |

**Frontends/domains**

| packt.azurefd.net                                           ⌄ |

**PATTERNS TO MATCH**

Set this to all the URL path patterns that this route will accept. For example, you can set this to /users/* to accept all requests on the URL www.contoso.com/users/*. Learn more

/*                                                            🗑

| /path |

Figure 13.12: Adding routing rule details

14. Under **ROUTE DETAILS**, we need to provide details for **Route type**, **Backend pool**, and **Forwarding protocol**. We can optionally select whether we want to enable the **URL rewrite** and **Caching** options:

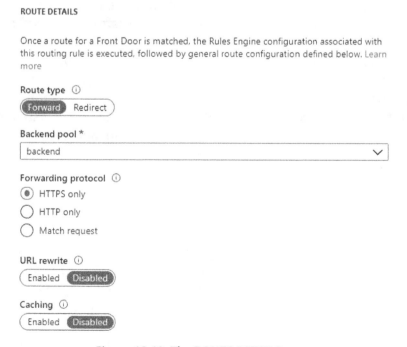

**ROUTE DETAILS**

Once a route for a Front Door is matched, the Rules Engine configuration associated with this routing rule is executed, followed by general route configuration defined below. Learn more

**Route type** ⓘ

( Forward ) Redirect

**Backend pool** *

| backend                                                     ⌄ |

**Forwarding protocol** ⓘ

◉ HTTPS only

○ HTTP only

○ Match request

**URL rewrite** ⓘ

( Enabled **Disabled** )

**Caching** ⓘ

( Enabled **Disabled** )

Figure 13.13: The ROUTE DETAILS pane

15. Once the routing rule has been created, we have all the necessary components and can proceed with creating the Azure Front Door instance by navigating to the **Review + create** tab:

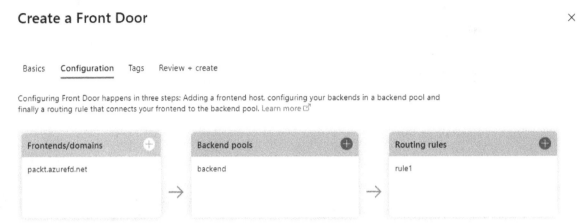

**Figure 13.14: All the components are configured**

## How it works...

All application requests are coming to the front end. Based on the rules we created, requests are forwarded to endpoints in the back end. Load balancing rules will ensure that requests will be sent to the fastest available back end.

The successful sample rate ensures that endpoints in the back end are available and determines how many samples are sent at a time. **Successful samples required** defines how many requests need to be successful in order for an endpoint to be considered healthy. **Latency sensitivity** sets the tolerance between the endpoint with the lowest latency and the rest of the endpoints. For example, let's say the **Latency sensitivity** setting is 30ms, while the latency of endpoint A is 15ms, that of endpoint B is 30ms, and that of endpoint C is 90ms. Endpoints A and B will be placed in the fastest pool as the difference in latency is lower than the sensitivity threshold, and endpoint C is out as it's above the threshold.

Routing rules define how traffic is handled and whether specific traffic needs to be redirected or forwarded. If **URL rewrite** is enabled, we can construct a URL that will be forwarded to a backend. If caching is enabled, Azure Front Door will cache static content for faster delivery.

> **Note**
>
> A lot of terms and options are the same as for Application Gateway and we will not explain them again. Furthermore, **Web Application Firewall** (**WAF**) is an option that can be enabled on Azure Front Door for better security. For more information on WAF, see the related recipes in *Chapter 12, Azure Application Gateway and Azure WAF*.

Azure Front Door also includes a number of configurable options and rules that can help your web applications deliver a customer- and brand-centric service. Here are some more important resources related to Azure Front Door:

- **Learn more about custom domains**: https://docs.microsoft.com/azure/frontdoor/front-door-custom-domain

- **Learn more about wildcard domains**: https://docs.microsoft.com/azure/frontdoor/front-door-wildcard-domain

- **Learn more about Rules Engine**: https://docs.microsoft.com/azure/frontdoor/front-door-rules-engine

- **Learn more about Rules Engine match conditions**: https://docs.microsoft.com/azure/frontdoor/front-door-rules-engine-match-conditions

- **Learn more about Rules Engine Actions**: https://docs.microsoft.com/azure/frontdoor/front-door-rules-engine-actions

Having created the Azure Front Door instance, let's move on to the next recipe and learn how to create an Azure CDN profile.

# Creating an Azure CDN profile

**Azure Content Delivery Network (Azure CDN)** is a distributed network that enables the faster delivery of web content to end users. Azure CDN stores cached content on edge servers in multiple locations (Azure regions). This content is then available to end users faster, with minimal network latency.

## Getting ready

Before you start, open the browser and go to the Azure portal via https://portal.azure.com.

## How to do it...

In order to create a new Azure CDN profile, take the following steps:

1.  In the Azure portal, select **Create a resource** and then choose **CDN** under **Networking** (or search for CDN in the search bar).

2.  In the new pane, we must provide information for the **Name**, **Subscription**, **Resource group**, and **Pricing tier** fields. If we decide to provide a CDN endpoint at this time, we need to provide details for **CDN endpoint name**, **Origin type**, and **Origin hostname**. **Origin hostname** will be available from the drop-down list, based on the **Origin type** option selected:

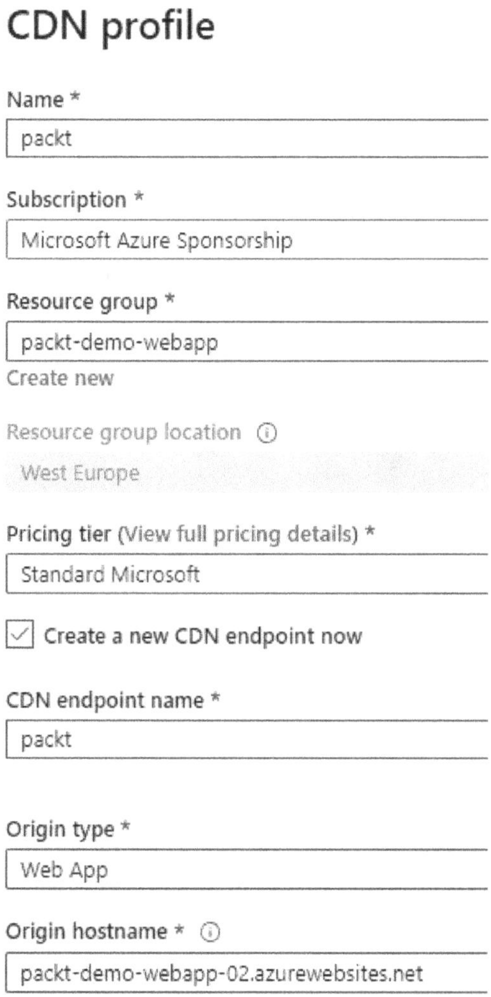

# CDN profile

Name *

packt

Subscription *

Microsoft Azure Sponsorship

Resource group *

packt-demo-webapp

Create new

Resource group location ⓘ

West Europe

Pricing tier (View full pricing details) *

Standard Microsoft

☑ Create a new CDN endpoint now

CDN endpoint name *

packt

Origin type *

Web App

Origin hostname * ⓘ

packt-demo-webapp-02.azurewebsites.net

Figure 13.15: Adding Azure CDN profile details

3. We can now create an Azure CDN profile. Following deployment, Azure CDN starts to cache the content of the origin and we can start using it immediately.

## How it works...

Azure CDN stores the content of our application on edge servers. As these edge servers are distributed across Azure regions, we have copies of content in practically every region in the world. Content is then delivered to end users from the closest location, which provides minimum network latency. Let's say an application is hosted in West Europe, and a user is located in the western part of the US. Content, in this case, will not be delivered from the original location, but from the location closest to the user, in this instance, West US. This way, we can ensure that each user has the best experience and delivery wherever they are.

# Index

**About**

All major keywords used in this book are captured alphabetically in this section. Each one is accompanied by the page number of where they appear.

## F

factor: 44, 67
failed: 237
failover: 206, 214, 250
faster: 259-260
feature: 240
features: 24, 250
filtering: 47, 123-124
firewall: 5, 72-73, 76, 78,
  82-83, 109-110, 112-116,
  118, 120-124, 170, 218,
  237-240, 242, 252, 259
firmware: 133
five-tuple: 192
floating: 191
forced: 109, 112, 118, 122
forcing: 152
format: 6, 8, 12, 102,
  123, 136, 205
forwarding: 192, 205,
  232, 235, 257
foundation: 1
framework: 6
frontdoor: 259
frontend: 5, 12, 51, 176,
  179-180, 182, 184,
  188, 190-194, 196-197,
  201, 205, 212, 214,
  217-218, 220, 223, 233
frontends: 220,
  222, 251, 256
functions: 134
further: 184, 207, 231
future: 1, 60, 227

## G

gateway: 10-11, 75-83,
  94-95, 102, 104-105,
  119, 122, 129-132,
  137-138, 140, 142-143,
  147, 152, 158-159, 161,
  164, 217-220, 226-230,
  232-234, 236-242,
  246-247, 250, 259
gateways: 75, 129-130,
  140, 143, 147, 155,
  227, 230, 234, 242
generate: 118, 134
generation: 79
geographic: 201, 209
global: 240, 244, 249-250
grayed: 70, 158, 164,
  176, 228, 231, 233,
  235, 237, 250
groups: 27, 33, 47-49,
  56, 109, 122-123
gwipconfig: 81

## H

handle: 213, 242
hands-on: 27, 188
header: 205, 253
headers: 205
health: 175, 179,
  188-192, 194, 205,
  214, 236-237, 255
hosted: 86-87, 93, 261
hostname: 260
hybrid: 21, 116, 127-128, 160
ignored: 39

## I

impact: 42
impacted: 227
implicit: 191, 194
important: 6, 41, 44,
  111, 116, 118, 127,
  134, 152, 259
inbound: 21, 37, 39,
  41-42, 50-51, 56,
  110, 179, 192-194
incoming: 33, 50, 109,
  190-191, 201, 205,
  217, 227, 232, 240
incomplete: 230
increase: 13, 179, 207, 214,
  218, 229, 237, 240
individual: 28, 123,
  174, 205, 242
in-flight: 232
initial: 6, 13, 63
initiate: 166
inspected: 122, 240
inspection: 118
install: 6
installed: 6
instance: 20-21, 110,
  112, 114, 123, 150-151,
  153-154, 194, 196-197,
  219, 249-250,
  258-259, 261
integrate: 89, 174
interested: 27, 190
interface: 19, 27-31,
  33-34, 44-46, 57, 59,
  63, 65, 67, 69, 155, 182
internal: 127, 179-183,
  185, 188, 190-192
internet: 2, 26, 39, 41-42,
  53, 102-104, 119-120,
  122, 170, 182, 184, 194

## O

observe: 13
obsolete: 107
on-demand: 140
openvpn: 138
operation: 217
operations: 82, 128, 229
optimize: 218
optimized: 217, 227
optional: 56, 144, 158-159,
    161-162, 169, 192, 231
organize: 25
origin: 208-209, 260-261
outbound: 37, 39-41,
    109-110, 112, 176,
    179, 191, 194-197
outcome: 36, 56
outgoing: 33
output: 7
outside: 14, 104, 127
overcome: 72
overlap: 8, 11-12, 16, 138
override: 225, 230-232
overview: 91, 132, 212

## P

parameters: 6-7, 11, 35,
    41, 47, 50, 55-56, 76,
    78, 113, 240, 244
passed: 26
password: 21, 155
patterns: 256
peering: 128, 143-144,
    146-147, 155-156
perform: 46, 128,
    134, 218, 227
permit: 140
physical: 25, 249
policy: 118, 166,
    218, 242-247

potential: 127
powershell: 1, 6-7, 11-12,
    34-36, 41-42, 46,
    54, 56, 75-78, 81-82,
    109-110, 113-115,
    127-128, 134, 202
predefined: 133, 138
prefix: 12, 54, 72-73, 78,
    102-105, 120, 152
premium: 22
pre-shared: 166
priority: 38-41, 44, 46,
    50, 114-115, 199, 201,
    207-208, 245, 253
private: 2, 11, 26, 28,
    53-54, 63-71, 73, 75,
    85, 87-90, 136, 143,
    147, 149-150, 158,
    164, 170-177, 180, 182,
    184, 205, 220, 233
probes: 175, 179, 188-189,
    218, 225, 236, 255
process: 20, 26, 28-31,
    36, 56, 81, 118, 140, 184,
    204, 214, 227, 254
processing: 229
profile: 199-202,
    205-206, 208-209,
    211-212, 249, 259-261
project: 219
protocol: 38, 40, 50, 76,
    80, 95, 127, 131, 164,
    166, 188-193, 196,
    217, 223, 225, 231,
    233, 237, 255-257
-protocol: 41-42, 114-115
protocols: 191, 206,
    217-218, 227, 233
provider: 165
proximity: 25

public: 4, 21-22, 26, 28,
    53-65, 67, 69, 71-73,
    76, 78, 80, 82-83,
    88, 103, 112-114, 122,
    127, 132, 139, 149,
    153-155, 175, 179-180,
    182-185, 188, 190, 192,
    194-195, 201, 203, 205,
    213-214, 220, 233
published: 188

## Q

qualified: 86, 123, 200
queries: 85

## R

ranges: 50, 123
rebooted: 61
record: 61, 85-87, 89-94
redirect: 223, 230
redundant: 56
regions: 199, 201-202,
    205, 213, 250, 260-261
registered: 86, 90
remote: 127, 147
repository: 133
requests: 179, 182, 184, 188,
    192, 205, 214, 217-218,
    227, 230, 232-233,
    235-236, 240, 247, 258
required: 1, 19, 22, 34-35,
    47, 54-55, 67, 71, 76,
    79, 82, 86-87, 89, 110,
    121-122, 128, 150, 152,
    154, 180, 188, 200, 218,
    242, 247, 249, 256, 258
requisite: 81
reroute: 102